FORD CORTINA
Mk1

Osprey AutoHistory

FORD CORTINA Mk 1

1962-66; 1200, 1500, GT, Lotus

JONATHAN WOOD

WILLOW PUBLISHING (Magor), Barecroft Common, Magor, Newport, Gwent, NP6 3EB, United Kingdom.

A catalogue record for this book is available from the British Library.

ISBN 0 - 9512523 - 8 - 0

Cover design for this re-print by A. R. Doe, The Studio, Llandevaud, Newport, Gwent, NP6 2AE, United Kingdom.

Printed in England by Morgan's Technical Books Ltd, P.O. Box 5, Wotton-Under-Edge, Glos, GL12 7BY, United Kingdom.

First published in 1984 by Osprey Publishing Limited.

Willow Publishing (Magor) and Tommy Sandham acknowledge with gratitude the help given by Shaun Barrington of REED ILLUSTRATED BOOKS, and JONATHAN WOOD, the Author, in bringing this re-print to publication.

Most photographs are courtesy of the Ford Motor Company. Owners of other photographs are asked to contact Willow Publishing (Magor).

Contents

Foreword by Sir Terence Beckett, CBE

The story of the Ford Cortina is more than just a tale about a best selling motor car. It is a story about the creation of a totally new segment of the British car market and a car that set new standards for two decades or more becoming in fact part of the folklore of the sixties and seventies.

It was named after the 1960 Italian Winter Olympics village Cortina d'Ampezzo—a decision which in some Ford circles might have been a prophecy for a catastrophic slide down the bobsleigh run or a triumphant winning ski jump placing Ford way ahead of its rivals.

The answer is of course well known.

The reason for the success is explained faithfully in this book. The dedicated and painstaking work performed by the product planning and engineering staff, the ingenious theme to create a new range of car—a medium range car at a small car price—the assembling of the management team and so on.

The Cortina quickly became a best seller. It found particular favour with the then infant but fast growing company car market. It is a tribute to its reliability that these buyers came back again and again in greater and greater numbers as the Cortina developed through to Mark Four.

This was particularly significant in the seventies when the economic problems in Britain and in Europe began to bite. The careful planning many years before providing the essential profit return paid off when many

Far right *Terence Beckett, Product Planning general manager at the time of the Cortina's creation, pictured with a 1965 model year car outside Ford of Britain's new headquarters at Warley, Essex, opened in 1964.*

Following his departure as Product Planning chief in 1963, Beckett became Marketing staff manager and was subsequently appointed Ford of Britain's chairman in 1976, a post he held for four years. He was knighted in 1978 and has been director-general of the Confederation of British Industries since 1980

of Ford's rivals were left in real trouble.

This book is a fitting tribute to the many people in the Ford motor company who made the Cortina the most successful British car ever.

Terence Beckett
London,
November 1983

Introduction

It is a curious paradox that Ford, by far and away Britain's most successful car company for many years, has only recently attracted the attention of motoring historians. Maybe the very utilitarian nature of its products up to the 1950s has engendered a certain resistance, but the extraordinary transformation of the company's image with the arrival of the Cortina in the sixties, coupled with a highly successful competitions programme, have given Ford products an added momentum that continues to this day.

In setting down the evolution of the Archbishop project, which flowered as the Cortina in the autumn of 1962, I have laid some stress that the concept represented a very real competition between Dagenham and the American/German Cardinal destined to emerge as the Taunus 12M. That the Cortina looked, performed and sold better than the Taunus is an accurate reflection of the abilities of outstanding British planning and engineering teams, who despite a seemingly impossible schedule, saw the project through to ultimate success.

I have also not hesitated to draw parallels between the Cortina and its exact contemporary, the British Motor Corporation's Morris 1100. The technically advanced BMC car offered its customers reliable, economic transport coupled with the road holding advantages of front-wheel-drive and remained as Britain's top selling car for close on a decade. In fact it achieved practically everything expected of it except to generate sufficient profits for its manufacturers. The contrast to the Ford approach to car design with its tight-reined fiscal constraints, epitomized by the Cortina, provides its own lessons which BMC, later restructured as British Leyland, belatedly acknowledged with the arrival of its Morris Marina in 1971.

In addition, I trust what emerges from the Cortina story is the far sighted shrewdness and sagacity of Sir Patrick Hennessy, whose outstanding commitment to the running of Ford's British affairs in the fifties and early sixties has seldom achieved the recognition it undoubtedly deserves. His early introduction of Product Planning to Dagenham, a policy of graduate recruitment and an enlightened reinvestment strategy, has paid dividends for Ford in periods of both growth and retrenchment.

Hennessy's flair for the appointment of subordinates is mirrored in the success of the Archbishop project, and I can do no better than echo the sentiments of Hamish Orr-Ewing, a key member of the Archbishop Product Planning staff, when he recently wrote 'The Cortina was a marvellous example of what can be done in this country when we set our mind to it and are properly led'.

Jonathan Wood
Farnham, Surrey
January 1984

CHAPTER ONE

Ford in Britain

When Ford announced its Consul Cortina in the autumn of 1962 there were the inevitable adverse comparisons with the British Motor Corporation's rival sophisticated front-wheel-drive Morris 1100. But the Cortina—that epitomized Ford's established formula for conventional, no nonsense mechanics at a realistic price—was destined to become the fastest selling car in the history of the British motor industry and is arguably the country's most significant mass produced model of the postwar years.

So why is this deceptively orthodox car so important? First and foremost the Cortina was an exceptionally light and therefore cheap medium sized family saloon; the result of an impeccable exercise in terms of timing, weight specification and costing, and is thus an impressive endorsement of the Product Planning concept that Ford alone practised within the British motor industry at that time. For not only was the Cortina a very popular car it was also an extremely profitable one, and in the all important 1960s it played a significant role in the decline of BMC and its resultant take over by Leyland Motors. In international corporate terms it was conceived as a British answer to Ford's American/German Taunus 12M which it decisively outstripped in terms of specifications and sales.

Above all, the success of the Cortina represented a triumphant culmination of Sir Patrick Hennessy's stewardship of Ford of Britain's affairs during the postwar era, and to see just how this came about it is

Ford's British sales were built on the tough, spidery and utilitarian value for money Model T, built at Trafford Park, Manchester between 1911 and 1927

necessary for us to briefly retrace our steps to when Ford cars first made an appearance on our country's roads.

Within months of Henry Ford founding his Detroit based Ford Motor Company in 1903 his cars were on sale in Britain at the Central Motor Company's premises in London's Long Acre, established by an enthusiastic young salesman, Percival Perry. It wasn't until 1908 that Ford introduced his all conquering Model T and assembly began in Britain in 1911 at Trafford Park, which was the country's first industrial estate, near the Manchester Ship Canal. From 1913 the Model T was mass produced in Detroit and Britain's first moving track assembly line soon followed at the Manchester factory. In 1913 British Model T production stood at 7310 chassis making it the most productive car plant of the European motor industry. Production was main-

The Model Y 4-door saloon, introduced in 1932 and destined to give Ford a renewed grip in the British market it holds to this day. In 1935 the 2-door version was reduced in price from £115 to £100, the first British four-seater saloon to sell at this figure

tained during the First World War, after which Ford's output soared with no less than 46,362 Ts and TT trucks leaving Trafford Park in 1920. This made Ford easily the country's largest car manufacturer with Austin next in the league at only 4319 cars. The flood of value-for-money Fords was undoubtedly a factor in a horsepower tax being introduced in Britain at the beginning of 1921. This meant that a vehicle's Road Fund Licence was geared to the RAC rating of its engine and calculated on the size of its bore. So the owner of a Model T, rated at 22 hp, paid £22 a year while his next door neighbour, who bought one of the increasingly popular Morris Cowleys with its 11.9 hp RAC rating, £12 per annum; around half the T's rate. Although Ford continued to be the country's largest vehicle producer for a further two years, Morris forged ahead in 1924 and Trafford Park output dwindled. Model T production ceased in 1927 and was replaced by the Model A. During the following year

only 6685 chassis were built, the lowest Ford figure since 1912.

Henry Ford chose Dagenham for his principal European manufacturing plant mainly for its proximity to the Thames and thus good docking facilities. Opened in 1931, the factory later boasted its own power station (right) as this 1962 photograph shows

Despite this patchy performance, Henry Ford was anxious to have a purpose built British factory on the lines of his mighty American Rouge River plant. Although Percival Perry—who had run Ford's British operations until he clashed with Ford and departed in 1919—favoured Southampton, the company opted for a Thames-side site at Dagenham, Essex and in 1928 work began on the new factory. The intention was to mass produce the 14.9 hp Model A there, but the combined effects of the world depression and the fact that this very American product was too big for the British market, meant that when the first stage of the plant was opened in 1931 it had very little to build. Yet the highly integrated factory was the 'Detroit of Europe'; with its own power station, docking facilities and blast furnaces to manufacture its own steel—though in truth, the latter

Sir Patrick Hennessy (right), who so ably steered Ford's fortunes in the crucial postwar years, pictured with a 1965 model Cortina in America and John Dugdale, the Society of Motor Manufacturers and Traders' North American representative on his right

attributes only came into their own in the postwar years as vehicle output soared. There was no body shop, but the American Briggs company were encouraged to build a plant alongside the new factory. It is difficult to imagine a greater contrast to William Morris's facilities with a scatter of factories dotted around Oxford and the Midlands.

But despite a potential output of 200,000 vehicles per annum, Dagenham faced financial ruin unless it could manufacture a small car tailored to the needs of the British and European markets. In October 1931 Percival

The new generation of postwar Fords typified by the Consul for 1951. The full width bodywork, overhead valve, oversquare engine and MacPherson strut independent front suspension were introduced on this 4-cylinder car and the Zephyr six announced simultaneously. The location is Ann Hathaway's cottage, Stratford-upon-Avon

Perry—who had been wooed back to the Ford fold in 1928—was in Detroit making a case for a new model. The result was the hastily designed 8 hp Model Y which went into production at Dagenham in 1932. It proved an instant success and gave Ford a renewed grip on the British car market that it holds to this day. In 1934 the Y held a massive 54 per cent of the 8 hp market but Morris, the traditional leader, fought back with his own Eight in 1935. Perry countered with a cost-cutting £100 Y, the first British four-seater saloon to be offered at this price. In 1936 and 1937 Ford moved forward to become

Britain's largest vehicle producer, ahead of Morris, for in addition to cars it had a strong commercial vehicle line and a near monopoly in tractor output. The Eight was restyled for 1937 and a 10 hp derivative, along with two V8s, constituted the Ford model range. Although output dipped prior to the outbreak of the Second World War, Ford was well established as one of the British Big Three motor manufacturers with only Morris and Austin producing more cars. However, Ford's 1939 profits of £1.7 million were only about £140,000 behind Morris's and around three times those of Austin.

Ford's commercial vehicle output was maintained during the Second World War and by 1946, it once again became the country's largest vehicle producer; a position held until 1952 when the Austin and Morris companies combined in the face of the challenge to form the British Motor Corporation. During this period Ford also consistently outstripped its British rivals in profit terms. In 1950, for instance, it made £9.7 million, compared with Morris's £7.1 and Austin's £5.2. At this time Ford was producing cheap, value-for-money vehicles in the shape of the prewar Anglia and Prefect, with Model T inspired transverse leaf suspension and side valve engines, mechanical brakes, six volt electrics and vacuum windscreen wipers of legendary inefficiency. However, a new generation of Fords arrived for 1951 in the shape of the 4- and 6-cylinder Consul and Zephyr models with full width monocoque bodies, MacPherson strut independent front suspension and oversquare (the horsepower tax which penalized bore size having been finally abolished in 1948), overhead valve engines. The smaller similarly styled but side valve engined 100E followed later for 1954, but the cost conscious theme was maintained by the Anglia/Prefect derived Popular which was the cheapest car on the British market until its demise in 1959.

Sir Patrick Hennessy (1898–1981) was directing the company's affairs in these crucial postwar years. He had been appointed managing director in 1948 and

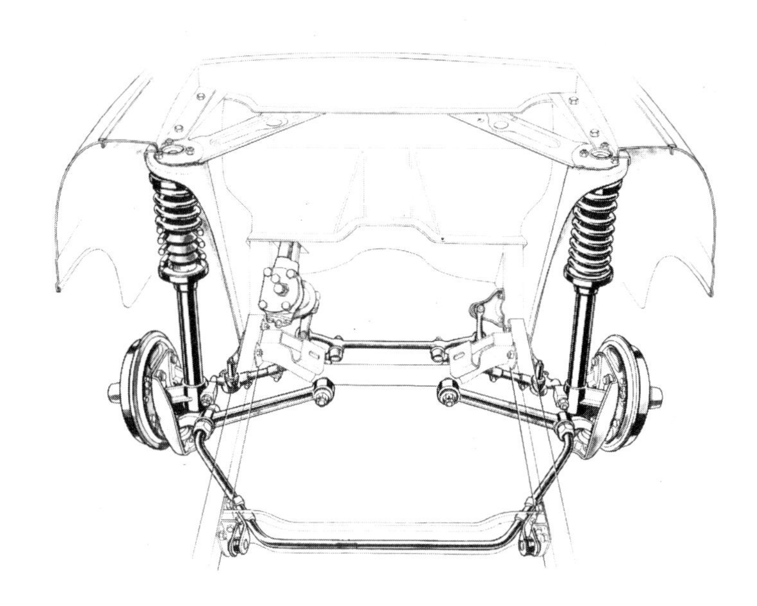

The MacPherson strut independent front suspension layout, introduced to Ford in Britain on the Consul/Zephyr range and adopted in considerably refined form on the Cortina

chairman in 1956, a position he held until his retirement in 1968. Hennessy was Irish born from Middleton, County Cork, and as a youngster had run away from home to join the British Army finally being commissioned in the Royal Inniskilling Fusiliers. After the war, in 1920, he joined the Henry Ford and Son tractor factory opened in his native Cork in 1919, for he was keen on sport and wanted to build himself up for confrontation on the rugger field! His intuitive, quick minded shrewdness soon made its mark and in 1931 Hennessy moved to Dagenham where he came to the attention of 'Cast Iron' Charlie Sorensen, who was Ford's Detroit based production czar and, like Henry Ford and his son Edsel, was a director of their British subsidiary. Hennessy was made purchasing manager at Dagenham, a department where the fiscal disciplines produced no less than three postwar chairmen: Hennessy himself, his predecessor Sir Rowland Smith (1950–56) and his successor Sir Leonard Crossland (1968–72). It fell to Hennessy to effect the economies that would reduce the price of the 1935 Model Y Popular by

£20 to £100. This was achieved by scrupulously assessing every part and pressing the company's suppliers to reduce prices. As a result the £100 Y's share of the 8 hp market spiralled from 22 per cent in 1935 to 41 per cent the following year. Hennessy was on his way. He visited America and established a good rapport with the Fords father and son, where his Irish brogue may have reminded Henry that his own grandfather had left County Cork for America back in 1847. Patrick Hennessy was made general manager at Dagenham in 1939 and soon after the outbreak of the Second World War in the dark days of 1940, he was seconded by Lord Beaverbrook to the Ministry of Aircraft Production. Hennessy became Beaverbrook's right hand during his time at the MAP and he received a knighthood in 1941 for his services to the war effort. At the end of the war, in 1945, he was once again able to devote himself to corporate affairs and became a director of the Ford Motor Company. 1948 saw Sir Patrick Hennessy take over the managing directorship from Sir Rowland Smith in the same year that Ford's British company inched ahead of its Canadian subsidiary to become the most productive element in its corporate global empire, outside America. In 1953 an important consolidation of resources occured, when Hennessy succeeded in purchasing Briggs Motor Bodies as he feared that Chrysler would buy the American parent (as indeed they did) and then control Dagenham's body supply. He later spoke of the event as one of the most important in Ford England's postwar history. By this time Sir Patrick Hennessy 'by wide reading, travel and friendships . . . made himself a cultivated, urbane gentleman' wrote Nevins and Hill in *Ford: Decline and Rebirth*, the definitive corporate history.

Meanwhile things had been far from well in Detroit, for in 1945 Ford was an ailing giant haemorrhaging funds at the rate of $10 million a month. Edsel Ford had died in 1943 at the tragically early age of 49, so Henry resumed the company's presidency. But in September

The spirit of the Model T lives! The 103E Popular, 1172 cc side valve powered with T inspired transverse leaf suspension, mechanical brakes and six volt electrics soldiered on until 1959 as the cheapest car on the British market. A total of 155,677 were built

1945, his grandson, Henry Ford II, took over the corporate reins, and aided by the legendary top management whizz kids, implemented a sweeping range of much needed reforms. This included replacing a system of model selection that had relied on hunch and whim, with a carefully orchestrated programme of Product Planning. Ford recognized that a firm's chief executive could no longer be responsible for only one or two models, as had been the case, but for as many as six vehicle projects. This delegation of responsibility was vested in the Product Planners and their Red Book, more of this anon.

In Britain in 1950, Sir Patrick Hennessy, with some prompting by Detroit, initiated a policy of graduate recruitment to the British Ford company. Previously it was rare to find university-educated engineers and managers in the British motor industry, though Rolls-Royce, Riley and Daimler were some of the companies that had a small graduate intake during the inter-war years. Some manufacturers positively discriminated against them. 'The university mind is a hindrance rather than a help' opined Sir Herbert Austin in 1930, sentiments that were then echoed by Ford's own Percival Perry. Lord Nuffield, as Sir William Morris became in 1934, was positively hostile to graduates (as is sometimes the way with self-made men) and there were instances of luckless degree holders getting the sack. This view still held when the Austin and Morris companies combined to form the British Motor Corporation in 1952. A policy of graduate recruitment was only introduced after its chairman Sir Leonard Lord, who shared the founding fathers prejudices, retired in 1961. Even then it was only implemented in a half-hearted way—in 1967 only one graduate joined the corporation.

Back in 1950 though it was Ford's enlightened and progressive policy that attracted the attention of 27 year old Terence (Terry) Beckett, who had just graduated with a BSc (Econ.) from the London School of Economics. Beckett, who had been educated at Wolverhampton Municipal Secondary School, wanted to be an economist but these ambitions were cut short by the outbreak of the Second World War. He therefore took an engineering course at the Wolverhampton and South Staffordshire Technical College qualifying as an AMI Mech E. Then followed national service with the Royal Electrical and Mechanical Engineers from whom Beckett was demobbed in 1948 with the rank of captain. He was then in a position to pursue his original career objective at the LSE. Having obtained his degree, Beckett looked around for an employer who was offering

The first effects of Ford's new Product Planning department were felt on the Zodiac Mark 11 (206E) of 1956. 80,000 were made though the cheaper Zephyr version found 221,000 new owners

a management training scheme for an engineer *and* an economist. As it happened there weren't many but Ford was a notable exception so, in 1950, Terry Beckett joined the Ford Motor Company at Dagenham on an 18 month management training scheme. Fortunately this initiation period was reduced to a year as the new recruit's outstanding abilities were recognized by Ford's management and in 1951, 27 year old Beckett became Sir Patrick Hennessy's personal assistant. There he looked after the product and engineering side of the business for Hennessy, later became styling manager and in 1954 he was put in charge of body engineering. The following year Beckett's meteoric corporate rise took a major step forward when he was appointed manager of Ford's Product Planning staff. At 31 he was the youngest divisional manager in the company's history.

It was in 1953 that Sir Patrick Hennessy, in the face of considerable opposition at Dagenham, introduced the concept of Product Planning to Ford of Britain. Already established practice within the corporate trans-Atlantic parent, its first manager in Britain was Martin Tustin (who had joined Ford in 1933 and came to the job after being head of the firm's parts division). Philip Ives, who became a member of the Product Planning staff in 1954, prior to Beckett's arrival, recalls 'there were about six people there when I joined; we were in a temporary building next to the sports club at Dagenham and used to have lunch in the pavilion!'. Tustin remained head of the division until he left at the end of 1954 to become general manager of Standard-Triumph where he played a leading role in the development of the Triumph Herald.

On his arrival Beckett created three departments concerned with light, medium and large cars, and two for commercial vehicles. In addition there was a separate advanced programme unit which looked more than five years ahead, and was also important as far as the company's long term engine plans were concerned. In the small car sector, work was begun on what was to be the 105E Anglia, and a new research and development establishment was opened in Lodge Road, Birmingham to tap engineering talent in the Midlands. Despite Hennessy's policy of graduate recruitment only four personnel with engineering degrees joined Ford in 1955 though the commercial side fared rather better. In addition, work was advanced on revamping the Consul/Zephyr range which appeared in Mark II form in 1956, to be followed by completely new Mark III versions in 1962.

Both these vehicles ranges were developments of existing lines but there was also a gap for a medium sized

car between the two, in a market that was at the time dominated by the Hillman Minx and its derivatives. The resulting new model is of considerable relevance to our story because the Classic, as it was eventually titled, represented a deliberate attempt by Ford to move away from its traditional, low cost, value-for-money role and produce a more expensive car, the sort of vehicle a man might take to his golf club without embarrassment. The new car, designated 109E, was intended to appear ahead of the Anglia, but in the end the latter was launched first

The 105E Anglia, with its distinctive inclined rear window, appeared at the 1959 Motor Show where it was £93 more expensive in its most basic form than BMC's front-wheel-drive Mini. The Anglia, a completely new car with no carry overs from previous models, was powered by a 997 cc oversquare overhead valve engine. This is a 1963 Super with 1200 engine

because of the urgent need to replace the side valve engined 100E.

While the company's engineers were concentrating their efforts on the next generation of Ford cars and commercials, work was also in hand on a five year development plan. Initiated in 1954, the £75 million programme was aimed at increasing the Dagenham floor space by half as much again of which the principle feature was to be a new Paint, Trim and Assembly building—this was opened in 1959. Significantly the expansion was funded by reinvesting profits and even before it got underway, Ford's British assets were worth over £60 million—several millions greater than the British Motor Corporation's. Yet further development was the creation of a new parts depot at Aveley, about eight miles from Dagenham and it was to there that Product Planning moved in 1961.

The 105E Anglia made its appearance at the 1959 Motor Show and, with its distinctive inclined rear window and new overhead valve engine, it proved at the time to be the most successful model in the company's history. Yet despite this success, Ford was faced with a significant challenge from BMC, for also making its show debut in 1959 was the sensational front-wheel-drive, transverse engined, all independent suspension Mini—the most technically significant car in the history of the British motor industry.

Of greater concern to Ford was the new car's price. For the Mini in its most basic form sold for £496 which was £93 *less* than the equivalent Anglia, while in De Luxe form the BMC baby cost £536 which undercut the De Luxe 105E by £74. At a stroke BMC had undermined Ford's historical role of offering the lowest priced car on the market, the archaic £419 Popular having been discontinued with the arrival of the 105E.

Yet front-wheel-drive was more expensive to manufacture than the orthodox front engine/rear-wheel-drive layout characterized by the Anglia. How had this apparent manufacturing miracle been accomplished?

The world famous Mini, which undermined Ford's traditional role for producing the cheapest car in the motoring market place, selling for £496 in its most basic form. This 1959 example is the Austin Seven version named in memory of Longbridge's most famous baby car

Ford bought a Mini, stripped it down to its spot welds and subjected it to a minute cost analysis. The results of this investigation were astonishing for the team found that on their costings, which they knew to be superior to BMC's, the Corporation was losing around £30 on every Mini it sold, 'though I could see ways in which we could have taken costs out of it without in any way reducing its sales appeal' recalls Beckett. They concluded that the car was not so much over engineered but its materials were over specified; in short they were too good for such a cheap car. Furthermore Ford concluded that if the car had been priced at £30 more it would hardly have affected the model's sales. These conclusions were subsequently confirmed by close scrutiny of BMC's published accounts.

How could this extraordinary state of affairs come to be? Like practically all British car manufacturers at the time, BMC had no Product Planning department but relied on its chairman, in the Corporation's case Sir Leonard Lord, along with chief engineer Alec Issigonis, for initiating its new generation of cars. Lord had a reputation as one of the industry's finest production engineers, and Issigonis had been responsible for the much loved Morris Minor which was the first British car to sell a million. In 1952 he left Cowley to join Alvis, a few months after he had finished the designs of an experimental front-wheel-drive Minor, though the actual car wasn't completed until after he had departed. Early in 1956 Alec Issigonis rejoined BMC and set to work designing what was to become the world famous Mini. He was a self-confessed small car enthusiast, and had a high regard for Citroen front-wheel-drive expertise being greatly attracted by the configeration's space saving attributes.

The Mini of 1959 represented a major shift of engineering policy by BMC, for although front-wheel-drive cars enjoyed considerable popularity in continental Europe, British manufacturers had largely fought shy of the layout. Roadholding and packaging undoubtedly represented pluses for potential customers but on the debit side there were higher manufacturing costs which were not reflected in the Mini's astonishing £496 price. There is only one inescapable conclusion to be drawn in the light of Ford's detailed researches on the subject: that the Mini was drastically underpriced because its makers lacked an effective realisation of its manufacturing costs. Certainly when he was running Austin, Lord's pricing policy attained a legendary empirical quality; for instance, when he decided on how much the A30 (Longbridge's rival to the Morris Minor) would cost thus: 'What's Morris's bloody figure—£515—right make ours £510'.

Although Lord announced a £10 million expansion programme in 1954, the same year as Ford's £75 million package, much of BMC's plant and tools, particularly at Cowley, were in urgent need of replacement. This was undoubtedly a factor in BMC averaging a profit of £35 per vehicle in 1956, compared with Ford's £45. By 1961 the differential had drastically widened with Ford having pushed its figure up to £53 whereas BMC, by then deeply committed to front-wheel-drive, was only making £6 10*s* on each car it produced.

CHAPTER TWO

Archbishop versus Cardinal

The thread of the Cortina story now moves across the Atlantic to the United States of America. During the fifties Ford had embarked on a furious sales battle with Chevrolet, flagship of the rival General Motors conglomerate. It was during this decade that the ill-fated Edsel car, and division, were launched which are reputed to have lost Ford around $350 million, a reminder that Product Planning is not a universal

The Taunus 12M, introduced in 1952 by Ford's German subsidiary, it was powered by a 1172 cc side valve engine and boasted MacPherson strut independent front suspension. This is the P1 version available between 1957 and 1959

panacea for all marketing ills. In addition to meeting traditional competition, American manufacturers were faced with a further unexpected challenge from foreign imports spearheaded by the unconventional, unlovely but utterly distinctive Volkswagen Beetle. Just two Beetles, or Bugs as the Americans called them, were registered in 1949 and the figure had risen to a mere 980 by 1953. In 1959 no less than 120,442 Bugs found American owners but demand didn't peak until 1968

when 423,008 were sold. Detroit's response was to produce what was called the 'compact', essentially a halfway house between the large American car and the small imported one, though to European eyes it was still a big saloon. To meet the challenge of the imported car head-on with an even smaller car was a different proposition; since the relatively small volumes involved even to capture Volkswagen sales would in Detroit terms hardly justify the necessary tooling costs.

The 12M received a much needed front end face lift with revised radiator grille for 1960. Though still side valve powered, the 1498 cc overhead valve unit from the former 15M was an optional fitment. It was this model that the Cardinal supplanted in 1962

However if the 'sub compact', as the concept was designated, could be produced on a global basis and follow in the wheel tracks of the Model T Ford as a world car, then the idea might have made more manufacturing sense. In 1957 therefore Ford asked its German and British subsidiaries to submit designs for such a vehicle. In Cologne, August Momberger—who had joined Ford from Borgward where he had been responsible for the front-wheel-drive Lloyd, Goliath and Hansa models—predictably came up with a V4 powered car that featured this drive configuration. By contrast, the British design followed a traditional in-line engine/rear-wheel-drive layout. The model was coded Cardinal, which is a small red breasted American bird rather like a British robin.

At the same time as the Cardinal studies were initiated Henry Ford gave the go-ahead for the company's first compact car; the 2.3 litre 6-cylinder Falcon and a strong seller it proved to be. Consequently the sub compact was down graded in priority and its announcement date was put forward to 1965. The idea was taken over by a group of Ford engineers; Jack Collins, Fred Bloom and Roy Lunn, who preferred the Cologne concept to the Dagenham one and the launch date was again revised, this time to 1962. The entire project was then taken over by yet another engineering group led by Bertil Andren and Jack Prendergast. The intention at this stage was to build the Cardinal at a plant in Louisville, Kentucky and also in Ford's Cologne factory. For Ford Germany would produce a 1.2 litre version along with a 1.5 litre variant. The latter, with its attendant trans-axle unit, would then be shipped to America where they would be implanted into bodyshells produced on a duplicate set of dies.

Ford's German presence dated from 1925 when the Model T was assembled in Berlin, though the Cologne factory wasn't built until 1931. Initially cars were assembled from imported parts but with the Nazis' rise to power in 1933, products had to be manufactured in Germany. Therefore from 1934 the Model Y was

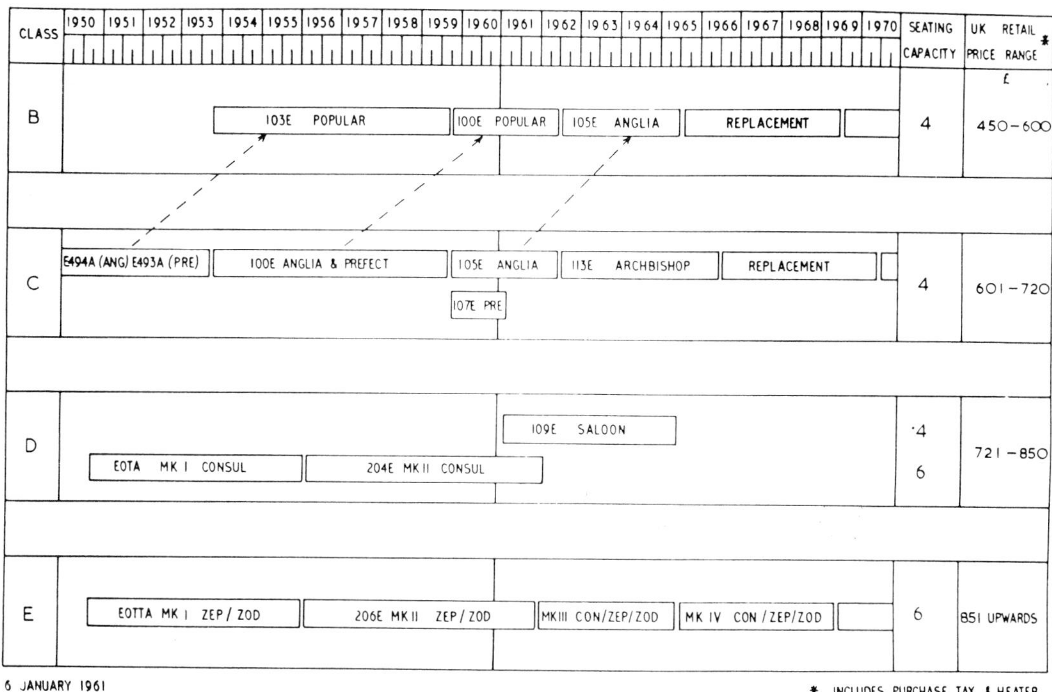

Ford's marketing strategy for the sixties, approved by the company's Product Committee in January 1961. Note that at this stage the Class C Archbishop (Cortina) was intended to be a replacement for the Anglia with the latter down priced for the Class B market

produced as the Köln. V8s followed, and 1939 saw the introduction of the Taunus, powered by a side valve engine similar to that used in the Dagenham built Prefect. The Second World War produced its inevitable upheavals after which Cologne just manufactured trucks until the Taunus was reintroduced in 1948. In 1952 came a new version, the 12M, which foreshadowed the British 100E of the following year and like its Dagenham counterpart was also 1172 cc side valve powered. It was joined by the overhead valve 1534 cc 15M in 1955 which, as with the smaller model, bore all the hall marks of trans-Atlantic origins having been conceived by German and American engineers in Detroit. The 17M followed in 1957 and was successfully revamped for 1961. The smaller car in the range—the

Fred Hart, who joined Ford in 1940, was appointed executive engineer, Light Cars, in 1959. He was responsible for the mechanical layout of the 105E Anglia and was to do likewise on the Cortina

12M—though partially restyled for 1960 was getting decidedly long in the tooth. Therefore it was intended that it be replaced by the new front-wheel-drive Cardinal which would also be marketed as the 12M. This would give Ford Werke a much needed shot in the arm because by the end of the fifties, despite an eightfold increase in output during the decade, it was still in fourth place in the sales league behind Volkswagen, Opel and Daimler-Benz.

When work on the Cardinal had been underway for about a year, the parent Ford company began to think in terms of a British involvement in the project. This would have marked a more formal integration of Ford's British and German operations. It was in the spring of 1960 that Sir Patrick Hennessy was advised of these thoughts: his response was swift and predictable. He summoned his Product Planning manager Terence Beckett and Fred Hart, executive engineer, Light Cars to his office to inform them of the American proposals and that he was determined to reject them. Hart had joined Ford in 1940, and postwar had worked on a 10 hp car project with Volkswagen inspired torsion bar front suspension. Although the design was stillborn, he was closely involved with the Consul/Zephyr and 100E range, and had been responsible for the overall mechanical layout of the 105E Anglia which was already proving to be a strong seller.

The difficulty was that work on the Cardinal had already been underway for a year and had a scheduled launch date of September 1962. Hennessy though was determined that Dagenham should produce a British response to the Cardinal, to echo its theme of a medium sized car at a small car price. Of course the problem was timing. It usually took over three years to develop a new model, and meeting the American/German challenge would mean that the launch date was a little over two and a half years away. 'I told Sir Patrick with that time factor nothing but a conventional layout would be possible' recalls Hart. But Ford were acutely aware that

Hamish Orr-Ewing, who was Light Car Product Planning Manager from 1959 to 1963, pictured at the Cortina's corporate launch in Paris in the summer of 1962. He left Ford for Leyland Motors in 1963 and joined Rank Xerox two years later; he has been its chairman since 1980

their new car would be selling against BMC's technically advanced front-wheel-drive Morris 1100 (which was then under development) and it was decided to produce an independent rear suspension system for the new car. 'It was again a matter of that deadline that meant we had to go to Ford in Detroit for the system which was developed there by Fred Bloom', Hart adds. A conventional live rear-axle/leaf-spring layout was designed in parallel.

One thing that the new project required at its most formative stage was a code name, Beckett came up with Archbishop in response to the American Cardinal. 'I knew that theirs was named after a bird but decided to call our car Archbishop as a bit of a joke. Our American colleagues never ceased pointing out that their name was not intended to be ecclesiastic!'

Fortunately some preliminary studies for a medium sized saloon had already been considered by Ford's Product Planning and engineering staff. When the 105E Anglia appeared in 1959 it proved to be a highly successful model but was only available in 2-door form.

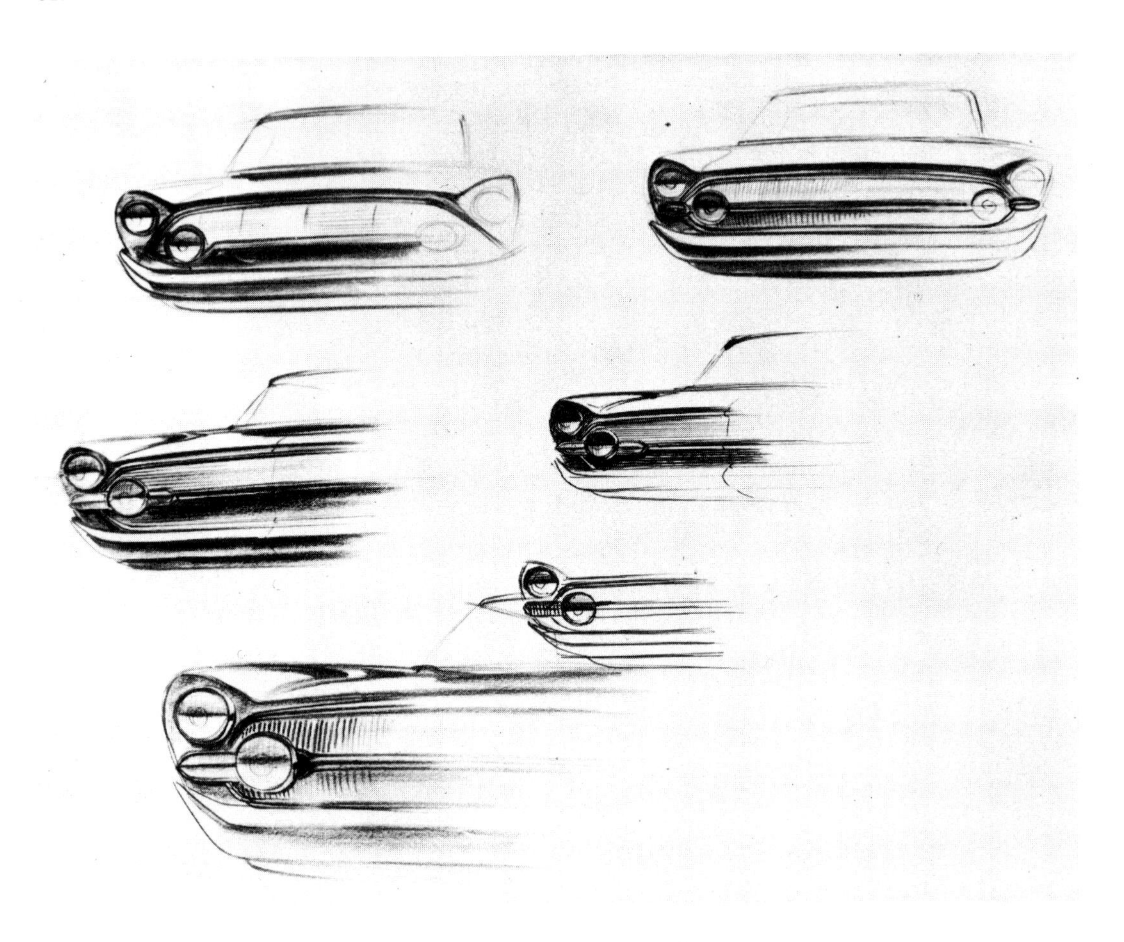

Styling sketches for the headlamp treatment on the Cortina (above) *which have more in common with the Anglia's front end which also appears to have been the inspiration for the study* (right*)*

However, its new overhead valve engine was also used in the old 100E bodyshell and the resulting 4-door 107E perpetuated this option until 1961. Therefore work was begun on what was then called the 4-door Prefect which would have used the 105E engine and drive train in a completely new 4-door bodyshell. Although the arrival of the Archbishop curtailed this activity, the studies already completed were not wasted. Similarly, after completing the 105E design, Fred Hart and his staff spent about three months carrying out some design work on a medium sized family saloon and again, these preliminaries were considered when work began in earnest on the Cortina.

There is yet a further dimension to the Archbishop project, for Beckett saw that it would represent an excellent response to the BMC Mini which was then achieving considerable sales success. To see just how this was attained, it is necessary to be aware of just how Ford divided up the car market when planning its future products. The divisions were as follows (but Class A represents bubble cars and three-wheelers and is thus outside the scope of this study):

Class	*Price*
Class B	£450–600
Class C	£601–720
Class D	£721–850
Class E	£851 upwards

The Mini was dominating Class B, and the 105E Anglia was a typical Class C product, while the Hillman Minx/Vauxhall Victor characterized Class D, and Ford's Zephyr/Zodiac range was Dagenham's Class E offering. What was to become the Cortina actually began life as an Anglia replacement, the intention being that the 105E be reduced in price to fall within Class B on the new model's announcement. But the concept of the Archbishop with its ingenious theme of a medium sized car at a small car price resulted in the creation of a completely new C/D car which would generate both high

Getting there; although the bonnet and radiator have yet to evolve, the headlamp treatment can be seen in the finished product while the all-important side flash is already present

volumes *and* profits. For in the higher class there were good profits but volumes fell off while in the lower there was a high volume potential but with a diminishing financial return, which was the reason that the Mini was proving to be a fiscal disaster for BMC.

While Fred Hart and Terence Beckett had more or less resolved the Archbishop's conventional but proven mechanics, the key to the car's viability rested with the new and, above all, light bodyshell. For if the low selling price was to be maintained it was essential that as much weight as possible be removed from the Archbishop's hull. Responsibility for this development fell on the capable shoulders of Don Ward, Ford's chief body engineer, and his colleague Andy Cox. Ward was an Australian and had come to Britain, via Ford of Canada, in 1940 when he had joined Briggs Motor Bodies at

A drawing of the Archbishop's rear with the distinctive sculptured boot lid already discernible. The circular 'CND' rear lights came later

Dagenham. It was here he earned a high reputation as a body engineer but after the war he returned to the antipodes. In 1958 he was wooed back and rejoined Ford of Britain. From 1960 onwards he set about seeking ways of producing a cheap, light bodyshell for the Archbishop. It was, he said, a matter of 'eliminating the unwanted passenger'.

In developing the Archbishop concept, the key considerations were cost, investment cost and weight. It was obviously not practical to apply these disciplines to all the car's parts, but to the key 500 as the remaining ones were usually co-related to them. This was attained by adopting a policy of triangulation. First the part from an existing Ford, in this case the 105E Anglia, was chosen and then compared with ones from competitors' models. In this instance those from the Hillman Minx

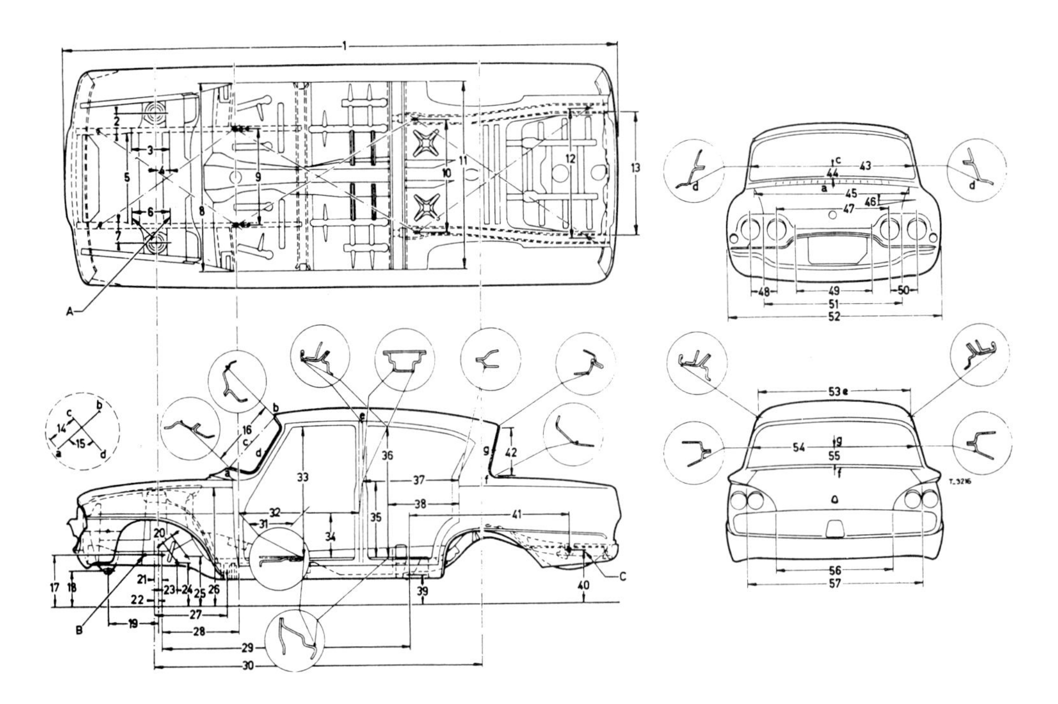

Contrast the relative simplicity of the Cortina's body structure (right) *with that of the Classic. The former was considerably lighter, had fewer components and was the result of aircraft stressing techniques being applied to the structure. Note the apparent absence of a boot floor on the Cortina. To save weight the top of the petrol tank did double duty as such*

and Volkswagen Beetle were employed. The cost and weight of the new part was then considered in relation to the other two. Beckett hand-picked a team of five executives to sign for the parts in question in the Red Book; an integral part of the Product Planning process that also hailed from America. This meant that each man was responsible for the part's specifications and that it was delivered on time; thus an element of responsbility was devolved from the chief executive. These requirements were then handed over to the engineering department, from where drawings were not released unless the demanding objectives had been achieved. Working in conjunction with the engineers were a team of 200 technical cost estimators who ensured that the financial demands were met. On occasions this proved to be impossible, but if a part breached the rigorous price ceiling then savings had to be made elsewhere to compensate.

Hamish Orr-Ewing was Product Planning Manager,

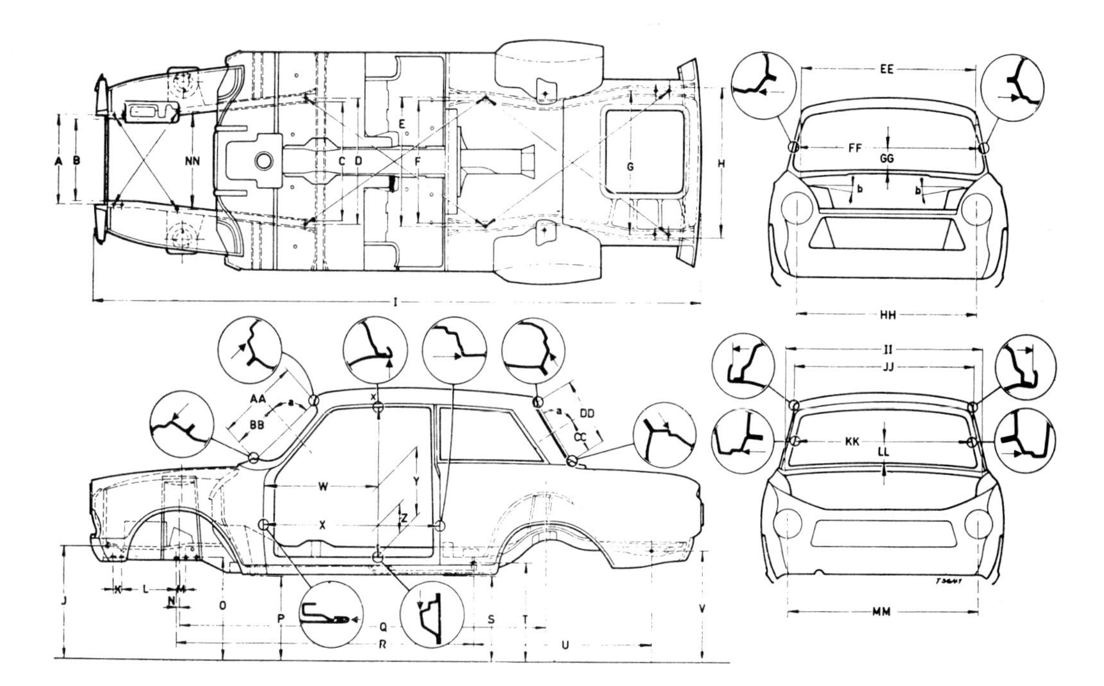

Light Cars who had direct responsibility for the Archbishop project. He is an Old Etonian, and prior to joining Ford in 1954 had been involved in his own modest car project with John Griffith producing the Ford V8 engined JAG from premises in Thames Ditton, Surrey. After a spell as a salesman with EMI, Orr-Ewing joined Ford's advertizing department and rose rapidly within the company to be appointed to his Product Planning post in 1959. He underlines the catalytic role played by the Cardinal project on the Archbishop team. 'Where we did a good job was in establishing objectives with the engineers which were extremely tough but just attainable. It was felt that if the Americans could do it why couldn't we.'

Just one of the Cardinal's less attainable objectives was a dry weight of 1630 lb (14.5 cwt) which was well known to Dennis Roberts, one of the key members of Don Ward's body design team. Roberts, who had gained a Ford Scholarship, joined the company in 1950 but after

Front end simplicity: the Cortina's engine and MacPherson strut suspension. This is the GT 1500 unit with twin choke Weber and four bladed fan. In its most basic form the 1200 model had a Solex carburettor and two bladed fan

completing his national service went to work in the London design office of the Bristol Aeroplane Company. He remained there for just over a year and a half, and in 1957 returned to Briggs Motor Bodies (which did not become fully integrated with Ford until 1958). Don Ward had recently set up a structures laboratory and wanted an engineer who was familiar with the theoretical side of body construction. 'I had this unique background of being trained as an automobile engineer and an aircraft stressman' adds Roberts.

It was mid-1960 when he got what was the first inkling of the Cortina project—Don Ward asked him to take the existing Classic structure, as a starting point of a new design, and remove around 150 lbs of metal in the process. 'The Classic was an uncontrolled design, there was nothing scientific about it'. He therefore applied the disciplines of aircraft stressing techniques to the Archbishop's structure, and at the same time drastically reduced the number of component parts so that eventually the Cortina had around 20 per cent fewer than the Classic. With this knowledge he began removing metal while following the stressman's axiom of 'if it bends it won't break', he reduced the depth of the car's underbody members and strove to eliminate the 'belt and bracers' approach that had characterized the Classic. Yet another feature of the Archbishop was that the top of the petrol tank did double duty as the boot floor rather than being placed underneath it, thus saving a few more precious pounds. In addition, Roberts also adopted the same informed approach to spot-welding and eventually was able to reduce the number of spots in the Cortina's upper structure by about 80.

An essential ingredient of the programme was, of course, the Archbishop's overall dimensions. Beckett

Rear end simplicity: the Cortina's rear axle, well forward of the centre spring line, intended to prevent plunge in the propeller shaft's sliding joint

and two colleagues spent around three weeks surrounded by seating bucks and pieces of cardboard finalizing the package. These were crucial considerations since the dimensions had to relate to the all important medium sized car specification with cost as an ever present constraint. An overall length of 14 ft 3 in., an unladen height of 4 ft $8\frac{1}{4}$ in. and a 5 ft $2\frac{1}{2}$ in. width was finalized, and it was then that the rival, smaller Cardinal was increased in size to bring it into line with the Archbishop. One up to Dagenham!

With the dimensions decided and work on the car's structure proceeding, it was then a matter of styling and these requirements filled eight detailed pages of typescript. Such demands as the tumblehome of the windows, height and contour of the roof and the

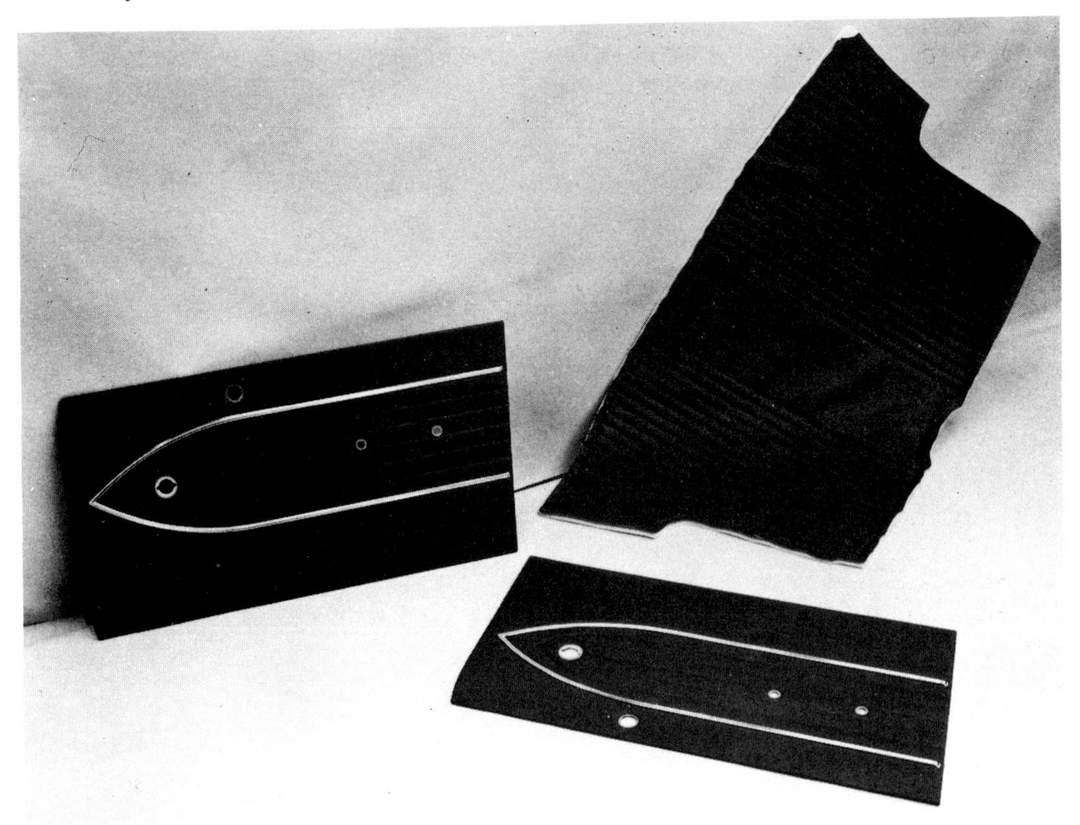

Examples of the Cortina's interior trim. Even these relatively minor items were meticulously costed

bonnet's power dome were clearly set down at this formative stage. A shift away from previous practice came with the abandonment of the inclined rear window introduced on the Anglia and perpetuated on the Classic. Although a car was cheaper to build with this feature, it produced a foreshortening effect, so out went that distinctive profile on the Archbishop where the emphasis on size was of crucial importance.

These detailed specifications were then presented to Roy Brown, Ford's chief stylist. Brown, a Canadian by birth and one time crooner, had joined General Motors' styling department in Detroit in the late forties. He remained there for only a short time before transferring to a design consultancy where he worked for American Airlines and Piper Aircraft. In 1953 he joined Ford where his first task was to design the Lincoln Futura dream car although he was subsequently responsible for the ill-fated Edsel range of 1958. In the following year Brown crossed the Atlantic and joined Ford in Britain, where he was soon at work on his first assignment—the estate car version of the 105E Anglia which appeared for the 1962 season. In addition to the Cortina and its estate variant, Brown also has the Mark III Zephyr/Zodiac range and Corsair to his credit.

Some of the Archbishop early styling sketches reveal unmistakable trans-Atlantic influences and Anglia related concepts. However, the finished product was a straightforward and functional design; a well proportioned saloon with a low waistline permitting a large window area. In all, five mock-ups were produced. One which particularly attracted Beckett's attention had a tapering flute along the side of the car, widening as it reached the vehicle's rear; this he considered to be the design's most single important feature. It created the necessary impression of length, broke up the mass of bodywork below the waistline while it also possessed a functional role by contributing to the stiffness of the body panels. The clay model was given approval in November 1960, a mere nine months or so since work had begun on the project. It was intended to produce the new car in three forms. The first to receive the go ahead was the 2-door version, with the 4-door variant being sanctioned in April 1961 and the estate version in September of the same year.

However there was one further change to be made to the Cortina's outward appearance which was also destined to become one of its most notable features. It was on January 27 1961 that John Bugas, head of Ford's International Division since 1959, visited Ford of Britain and viewed the Archbishop's projected body styling. A feature he objected to was the rear lights, which were of the dihedral type, as Detroit was by then committed to circular rear clusters. So new, round tripartite rear lights were designed and, although Mercedes-Benz

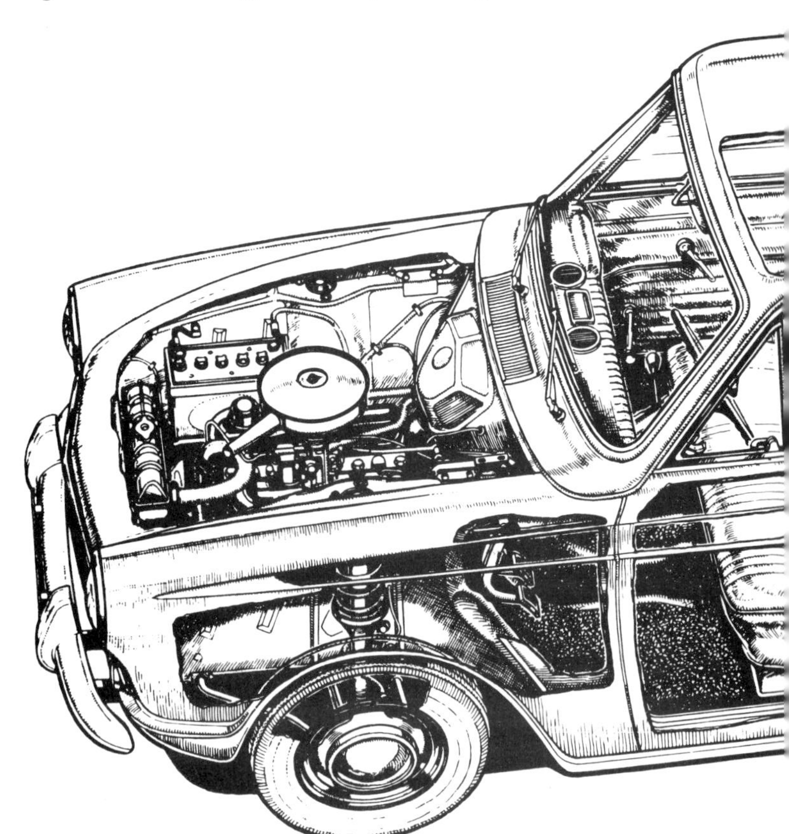

A triumph of convention: a Max Millar rendering of the 2-door De Luxe Cortina, no doubt drawn from one of the left-hand drive pre-production models ready for the model's corporate launch at Montlhéry track, near Paris

thought they might be too close to their own tri-star for comfort, the public instantly identified them, on the car's announcement, with the Campaign for Nuclear Disarmament motif! Beckett is convinced that the lamps are second only to that distinctive side flash in visual importance.

So just how did the Archbishop project affect the Classic, the medium sized saloon that Ford intended to introduce in 1960? Initially there was a delay in its appearance caused by the outstanding success of the 105E Anglia with a record then of 191,752 produced in 1960 (its full production year) which meant that the Classic's launch was delayed until May 1961. The Capri, a sporty variant, followed in September. Nevertheless the more that Beckett looked at the Classic the more he

676 VTW

became convinced that it was a heavy, high cost car which radically departed from Ford's traditional commitment of offering the best value-for-money model on the market; though the impending arrival of the Cortina would represent a positive restatement of these values.

Meanwhile in America the Cardinal was being subjected to one of its periodic reappraisals. 1960 had been a crucial year for the parent company as Ernest Breech, one of the principal architects of Ford's postwar renaissance, stepped down from the chairmanship. He was replaced by Robert McNamara, who as head of the Ford Division had just successfully launched the Falcon

Far left *The Consul Classic 315, introduced in 1961 but only produced until 1964. The first disc braked Ford, it was styled by Colin Neal, who was responsible for 105E Anglia's styling*

Below *The Classic's sporty derivative, the 2-door Capri of 1961. Initially 1340 cc powered, a 1500 cc engine was fitted from mid 1962 simultaneously with the Classic. In its GT form it shared similar mechanics to the Cortina GT*

Far right *The Archbishop becomes the Cortina. The location is the Ford International Marketing Conference at Montlhéry track near Paris. The Consul names qppears on the bonnet, a feature of the car until the 1965 model year*

Below *A De Luxe Cortina undergoing rig and static testing of its body structure in mid-1962. Note the model name taped over on the boot lid*

compact car. In the same year John Kennedy was elected American President, and in December he announced that his new administration's Secretary of Defence was to be McNamara, so Henry Ford himself took over as the company's chief executive. Robert McNamara's divisional job went to 36 year old Lee Iacocca who was convinced that Ford's strategy for the sixties should embody a strong sporting image. This theme was culminated in the arrival of the Mustang in 1964. Unfortunately this shift in emphasis didn't accord with the no frills pedestrian Cardinal, and Iacocca was later to claim that the best thing he ever did for Ford was to

CORTINA

The rival German/American Cardinal, announced as the Taunus 12M simultaneously with the Cortina in the autumn of 1962, pictured in Cologne where it was manufactured. It was a front-wheel-drive offering, powered by a 1.2-litre V4 engine but it was heavier, slower and was decisively outsold by the British Cortina

kill the car's American manufacturing base at the beginning of the youth boom. Henceforward the Cardinal would only be produced in Cologne where it was to be sold as the Taunus 12M, although work on the car continued apace in both America and Germany.

As already noted the rigorous timing demands were an important factor in the Archbishop having a conventional front engine/rear-wheel-drive layout. Yet a further crucial ingredient was that Ford already possessed what was to be subsequently known as the Kent engine range which had made its appearance in the 105E Anglia. (The series was not entitled Kent until the bowl in piston version arrived in 1967, and was so called because Alan Worters, Executive Engineer Power Units, lived in Kent! There already existed a V-engine

The Cortina and Taunus 12M meet at Montlhéry in June 1962. In addition to Fords, other rival models such as the Volkswagen Beetle and R4 and R8 Renault are also present

range called Essex which mirrored Dagenham's geographical location.)

The conception of the Kent family of engines provides a fascinating insight to Ford's rigorous cost conscious approach to all aspects of car design. It was back in 1954 that work began on a new generation of engines to replace the then current 1172 cc side valve unit which dated back to 1934. This old faithful weighed 223 lb and when work began on the new range the target aimed for was 190 lb, which proved impossible to attain though the 1200 cc version used in the Cortina weighed 220 lb. Although initially conceived for a 835 cc/1097 cc range this was later revised to 997 cc/1390 cc. When the engine was conceived, it was recognized as making the most effective use of transfer machinery: the cylinder block

TO

and head should be produced in volumes above 500 but below a 1000 units a day. As sales of the Anglia and Classic were predicted to be around 850 a day it was considered essential that as many common components were shared between the two units. In its original 835 cc/1097 cc form, a common stroke and different bore sizes were tried. However it proved cheaper to produce identical cylinder blocks with a 80.96 mm bore, and vary the specification of the crankshaft and connecting rods which was a relatively cheap and straightforward manufacturing exercise. Yet a further advantage was that a common bore meant a single sized piston, a great advantage to Ford's dealer organisation. An oversquare configuration was adopted, following on from the layout initiated on the 1951 Consul/Zephyr range where the advantages of lower piston speed and improved piston

Above *Henry Ford (*left*) sits next to Sir Patrick Hennessy, Ford of Britain's chairman, listening to the presentation of the Cortina's specifications in Paris, June 1962*

Far left *Henry Ford II (*centre*) examining the engine of the prototype Cortina GT at Montlhéry, while Fred Hart, executive engineer, Light Cars, points out an underbonnet detail*

ring and cylinder life proved an overwhelming attraction. For the Anglia's 997 cc an unprecedented 0.60 stroke/bore ratio was decided upon.

However a big bore demanded a long engine so height had to be kept to a minimum for consideration of weight, this was fixed at 7.125 in. as being suitable for both capacities. When it came to the crankshaft, Ford retained a three bearing unit and was also commited to the use of cast iron which was cheaper than a forging and required less machining; yet another manufacturing plus. The shaft and its webs were hollowed for lightness, this permitted rigidity and a large bearing area to be attained without extra weight—a manufacturing technique that had been perfected by Ford of Germany. This

Far left *A sectioned 2-door De Luxe at the model's Paris presentation, revealing the MacPherson strut independent front suspension*

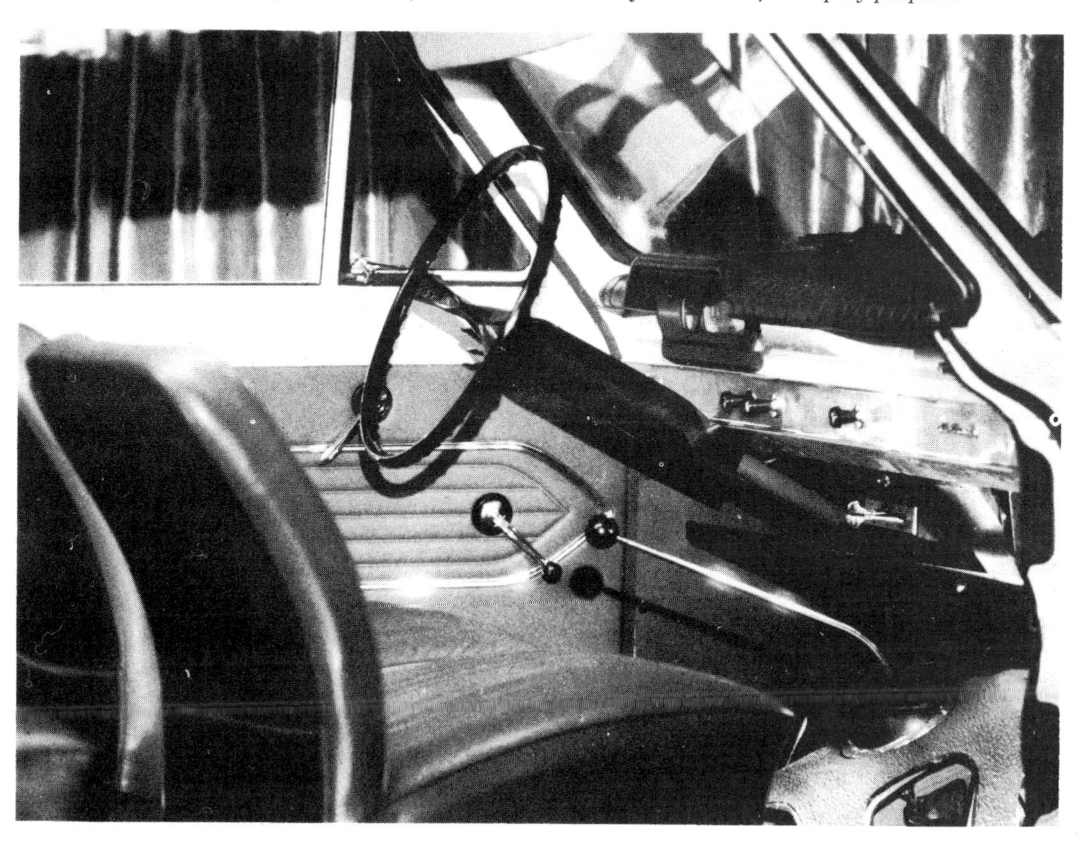

Below *Left-hand drive cars were featured at the corporate launch so that there could be accurate comparison with the Taunus 12M. This interior has been specially prepared for display purposes*

approach paid off: the shaft weighed 17.56 lb compared with the 1172 cc unit's 24.2 lb and that, of course, had smaller bearings. The overhead valve cylinder head, although common to both models, shared the same valves for ease of manufacture although the Anglia version didn't really require them. The oil pump and its adjoining filter was also a new layout and was externally mounted so as to speed up and cheapen the manufacturing process.

First on the scene was the 80 × 48 mm 997 cc Anglia unit. This was followed by a 1276 cc version for the weighty Classic announced in May 1961 but discarded in favour of a larger capacity, but noisier, 1340 cc version.

Ford executives giving the 4-door Cortina the once over in Paris, due to be revealed at the 1962 London Motor Show

It was decided that the Archbishop would use an engine between these two capacities and accordingly an 80 × 58 mm 1198 cc unit was developed. Therefore the only fresh components required were a new crankshaft, connecting rods and cylinder head of the correct combustion volume. As the shape of the bath tub type combustion chamber evolved the design and manufacturing engineers were in close co-operation so that they could be produced by four simple milling operations. The same engine was also used in the Anglia 1200 which was announced simultaneously with the Cortina.

However in 1960, prior to its announcement the following year, Ford realised that the Classic was going

The Super Estate version revealed at Paris in 1962 but not shown to the public until the Geneva Motor Show of the following year. The Di-Noc *simulated wood panel identify*

Far right *A 1962 Cortina undergoing gradient tests at the Fighting Vehicles Research and Development Establishment at Chobham, Surrey*

to require an even bigger engine, as its Hillman Minx rival was to be 1600 cc powered for 1962. This new engine retained the scared 80 mm bore, but the 72 mm stroke demanded that if the existing ratio was to be retained the height of the cylinder block would have to be raised by 0.62 in. This capacity increase also required a departure from the three bearing layout used hitherto since experiences with the 1340 cc version convinced the Ford engineers that the light block scantlings would result in a harsh, noisy power unit. This assumption was confirmed by building experimental 1498 cc engines with both three and five bearing crankshafts. As the loads were less on the five bearing engine it was possible to revert to micro babbit bearings which more than offset the cost of the additional shells. However, it was not possible to retain the hollow crankshaft featured on the three bearing units. This 1500 engine was announced for the Classic and Capri in August 1962, a month ahead of the Cortina. Ford also intended to use this larger engine in the export version of the Cortina, though it was subsequently decided to offer it on the home market in the Cortina Super (announced in January 1963).

The arrival of the 1500 engine in the Classic saw the simultaneous appearance of synchromesh on the bottom gear cog and it was this gearbox, with minor variations, that was used on the Cortina. There was a floor gearchange, though a steering column option was offered in conjunction with a bench type front seat, perpetuating Classic practice. Had the new car been aimed principally at the British market the floor provision would probably have sufficed. However Ford market research revealed that while 95 per cent of British customers preferred it, this only applied to 40 per cent of overseas ones. So the column facility was retained with the box being activated by British Wire Products cables. This permitted a straightforward change from right- to left-hand drive.

Superficially, the Archbishop's suspension of MacPherson struts at the front and a leaf sprung rear

N° 3
26½° 1 in 2
TOO 508

perpetuated a theme introduced on the Consul/Zephyr range of 1951 and was continued on the Anglia and Classic. However, the front struts used on the Archbishop were of a new type with the upper and lower bearings being more widely spaced than before to reduce stiction caused by cornering side forces. New damper fluid was introduced for the same reason. It wasn't until May 1961 that it was decided not to pursue the Detroit designed independent rear suspension system but to adopt a conventional cart sprung rear; though at this stage it was thought it might still be an option on the Super version of the Archbishop, or for the Classic replacement which emerged as the Corsair in 1963. So the rear suspension displayed advantages initiated on the Classic. The axle was mounted assymetrically on the springs but with $\frac{5}{12}$ of the spring forward of the rear axle. This was intended to reduce plunge in the propeller shaft sliding joint to the detriment of the rear suspension. The recirculating ball steering echoed Anglia and Classic practice while 8 in. Girling drum brakes were fitted all round.

The programme proceeded, right on schedule, and by the spring of 1962 about the only thing that the Archbishop required was a name. This had already caused Ford considerable problems. As early as March 1961 the company's Product Committee had decided that the new car would perpetuate the Prefect title. Although Anglia was also toyed with, by December there had been a change to Consul 225 for the 1200 and Consul 255 for the 1500 version. The numerical suffixes were purely superficial and are of no particular relevance. The Consul name was displayed on the dummy airscoop on the front of the Archbishop's bonnet to convey the impression of a large car. Then, at the last minute, Terry Beckett came up with the alliterative Cortina, as Cortina d'Ampezzo had been the Italian Alpine venue for the 1960 Winter Olympics and conveyed a fashionable, sporty image. It also gave the car a significantly European flavour, as at the time Britain was negotiating

GENTLEMEN - THE CONSUL CORTINA

THE CORTINA IS A 2 OR 4 DOOR CAR. IT HAS A 1200 cc 53.5 GROSS BHP ENGINE; A DEVELOPMENT FROM THE BRILLIANTLY SUCCESSFUL FAMILY OF ENGINES USED IN THE ANGLIA AND CLASSIC.

THE POWER IS TRANSMITTED TO THE REAR WHEELS THROUGH A NEW AND IMMENSELY TOUGH GEARBOX WITH SYNCHROMESH ON ALL FORWARD SPEEDS TO A 4.125:1 HYPOID AXLE COUPLED TO THE BODY BY LAMINATED SPRINGS AND DAMPED BY TELESCOPIC SHOCK ABSORBERS.

THE FRONT SUSPENSION IS A NEW, LIGHTENED AND LOWERED VERSION OF THE UNIQUE PILLAR SUSPENSION USED SO SUCCESSFULLY ON ALL FORD PASSENGER CARS.

THE STYLISH CORTINA IS A SMALL CAR WITH A BIG DIFFERENCE - WHY? - BECAUSE IT OFFERS THE INTERIOR SPACIOUSNESS AND OVERALL GOOD LOOKS ONLY FOUND UNTIL NOW IN CARS FAR MORE COSTLY TO BUY AND TO OPERATE.

LET'S LOOK AT SOME MEASUREMENTS -

THE NEW CORTINA HAS 13 cms MORE ROOM ACROSS THE FRONT SEATS THAN THE AVERAGE MEASUREMENT FOR SMALL CARS. - 23 cms ACROSS THE REAR SEAT. - 10 cms MORE FRONT LEGROOM - 7.6 cms MORE REAR LEGROOM AND - I WILL SAY IT VERY DISTINCTLY - NEARLY 2½ TIMES AS MUCH BOOT SPACE AS THE AVERAGE SMALL CAR ON THE ROADS TODAY.

Historic document: the first page of Hamish Orr-Ewing's speech to the motoring press on the occasion of the Cortina launch. Note the presentation of 'the medium sized car at a small car price' theme

for Common Market entry (though these overtures were vetoed by the French in 1963). In addition Cortina had the advantage of being a place name and consequently was not subject to the usual problems that can haunt model titles. Thus the Archbishop became the Consul Cortina.

The sporting ingredient was important as Ford were only too aware that while they had produced a roomy family saloon that would also appeal to the all-important fleet markets it was, as Beckett puts it, 'in overall specifications a rather ordinary motor car'. So in the spring of 1962 Sir Patrick Hennessy and Terence Beckett called in Colin Chapman 'and told him what we were doing to see if he could sprinkle a little star dust' on the new model. Just how Chapman achieved his objective and the conception and evolution of the Dagenham inspired GT are set down in Chapter Four.

A vital element, as yet unresolved, was that of the car's pricing. The Classic had failed to come up to sales expectations and it was generally accepted at Dagenham that one factor had been its relatively high price—£796 on its announcement in 4-door form, which was £70 more than the most expensive Hillman Minx. So the deliberations by finance director John Barber's staff were particularly crucial. After some three months of consideration they decided to price the Cortina according to what the market could stand, with the impending announcment of BMC Morris 1100 an ever present consideration. In fact the final prices ensured maximum returns for Ford, with the Cortina range destined to contribute around half the company's car line profits for many years to come. It was pitched between the Anglia and Classic, thus creating a fourth product line, as the related 1962 prices reveal:

Anglia De Luxe	£612
Cortina Standard 2-door	£636
Cortina De Luxe 4-door	£687
Classic 2-door	£723

The first production Cortina, Job Number One, went down the Dagenham production line on June 4th 1962. Ford were so confident that they had a winner on their hands that production targets were upped by 50 per cent. The project also represented an endorsement of Product Planning, in that by the disciplines of the Red Book, Beckett and his team achieved a 16*s* saving on the car's costing with the Archbishop £50,000 *under* its £13 million investment target.

Then came the all-important corporate response to the Cortina. Along with the rival Cardinal, by then titled Taunus 12M, cars were assembled at Montlhery track near Paris for appraisal by Ford executives from Henry Ford downwards. Beckett was in no doubt about which car was the victor. 'We had the edge on them in terms of reliability and style, and it took them a long time to get on top of the front-wheel-drive concept while we had a proven drive train'. The production figures of the respective cars confirmed this for the Cortina and its Mark II successor consistently and decisively outsold the concurrent Taunus models.

Cortina Mark I (1962–66)	1,010,090	Taunus 12M P4 (1962–66)	680,206
Cortina Mark II (1966–70)	1,010,580	Taunus 12M P5 (1966–70)	668,187

All that remained was the public response. First came the press launch held at London's Grosvenor House Hotel. Ford produced a lavish promotional booklet for the occasion, featuring the Cortina and the colour photography of Erich Hartmann. Significantly, the publication highlighted key members of the Archbishop planning and engineering team. Previously the company's personnel had remained in discreet anonymity, but by 1962 Ford had a new director of public affairs—Walter Hayes, formerly editor of the *Sunday Dispatch*.

The Consul Cortina was announced to the public in September 1962 with extensive magazine and newspaper advertizing by the London Press Exchange, though television promotion was confined to the day of the car's launch; how times have changed. The model was scheduled to appear at the London Motor Show which opened on October 17th. There the public would be able to view the Cortina and its BMC Morris 1100 rival. The battle for the sixties was on.

CHAPTER THREE

Cortina triumphant

The near half a million people who attended the 47th Motor Show at Earls Court in 1962 were greeted by a bevy of new models. There was Lotus's Elan making its show debut, along with the more popular MGB. While Standard Triumph, absorbed by a thrusting Leyland Motors the previous year, was exhibiting its sporty Spitfire for the first time. However there was little doubt

Cortinas reaching the conclusion of their production cycle at the new Dagenham Paint, Trim and Assembly building but yet to receive their road wheels, with 4-door cars in the foreground and 2-door examples in the background

'Car of the Year' accolades submerged the Cortina, and rightly so. Here's a Super at the World Fair in America plastered with appropriate flags from countries where it was the winner

Right *Rare bird: Ford rarely builds prototype cars unless they are intended for production but this was the Saxon, a two-seater version of the Cortina, built in 1962. It was sent to Detroit for evaluation but no more was heard of the concept*

Below *The immaculate engine bay of a fully restored ex-Works Cortina rally car, KPU 383C. Notice the attention to detail, such as the lamp on the bulkhead, useful for night-time adjustments!*

The Cortina in its most basic form. The no frills Standard model appeared in 1962 and continued until September 1965. This is the 1965 model year version with Cortina bonnet badge

Above *The De Luxe Cortina, as updated for 1965, showed a greatly improved dashboard with new three spoked steering wheel and Aeroflow heating and ventilation system*

Above right *The Super, in 1965 guise, featured body side-length chrome strips and wheel trims as its major recognition points*

Below right *The Cortina GT did great things to the medium sized performance saloon market. They were mostly seen in white or red—this white one has red upholstery. Note the four doors and Aeroflow 'vents'*

GWC 252B

GWC 252B

Right *The Cortina was always a strong export seller. It contributed in August 1963 to Ford announcing a record 312,000 cars had been sold abroad during the previous 12 months, the highest ever shipped overseas by a British car manufacturer.*

Far right *The Cortina GT made a fine rally car. In 1964 the car won the touring car category on the Alpine (this event) without substantial modification as is seen on rally cars today*

Below *The Lotus Cortina benefitting from the 1965 model year updates. Lotus badges can be seen on the radiator grille and rear wing. The green flash also identifies.*

892 DOO
28

Right *The Lotus Cortina, on the other hand, was a more successful racing car. Here is Sir John Whitmore zipping along in fine style in a Team Lotus car during a 1965 British Touring Car Championship round. These cars had radius arm/leaf spring rear suspension and BRM developed engines*

Below *Lotus Cortinas were raced the world over, usually with great success. This one was driven by Peter Feistmann and Jak McLaughlin in America in 1965. Note right-hand drive*

that the stars of the show were two family saloons: the British Motor Corporation's Morris 1100 and the Archbishop, revealed at last as the Ford Cortina.

The following table sets down the state of the market in October 1962, showing the strong similarity in the two models' specifications—though from an engineering standpoint they were poles apart. The Hillman Minx and Vauxhall Victor are also included to show how successfully Ford achieved its 'large car at a low price' theme.

	Price	*Weight*	*Length*	*cc*	*0–60 mph**	*Top speed**
Ford Cortina †	£639	15.5 cwt	14ft 0.3in.	1198	22.5 sec	76.5
Morris 1100 †	£675	16.25 cwt	12ft 2.7in.	1098	22.2 sec	77.7
Hillman Minx	£702	19.9 cwt	13ft 6.5in.	1592	23.6 sec	78.5
Vauxhall Victor	£702	19.2 cwt	14ft 5.25in.	1508	23.4 sec	76.5

† 2-door version * *Autocar* figures

The 1100 typified BMC's new design philosophy of producing technically advanced cars that could be manufactured over a long period, rather than changing conventionally engineered models every four years as was Ford practice. By adopting a policy of what the Corporation termed engineering excellence, BMC aimed to avoid the heavy retooling costs demanded by regular model changes. Though on the debit side, the profit margin on the front wheel-drive cars was less than on the north/south engined rear-drive ones.

As can be seen from the above table the 1100's engine capacity was similar to the Cortina; there the resemblance ended for its mechanics were totally at odds with established Dagenham practice. The 1100, coded ADO 16 in BMC model parlance, was a development of the theme Alec Issigonis established with his sensational front-wheel-drive Mini Minor. The 1098 cc engine was the reliable and economical BMC A Series unit with a 64 × 83 mm bore and stroke, while Ford opted for a 80 × 58 mm oversquare layout. As with the Mini, to save space the 1100's engine was mounted transversely with its gearbox ingeniously contained within the sump. This also embodied the differential unit with the shafts then taking the drive, via rubber inner and Rzeppa constant velocity outer joints, to the front wheels. Front disc brakes were a standard fitment while the Ford at this stage only offered drums. The 1100's Hydrolastic suspension was completely new. It was an all-independent interconnected system conceived by Moulton Developments which seemed a world away from Ford proven front MacPherson strut and leaf spring layout. The Morris was about 1 ft 9 in. shorter than the Cortina though the latter was nearly 0.75 cwt lighter. This reflected the stimulus from the Cardinal project and the demanding disciplines of the Ford Product Planners and their Red Book. The 1100 was available in 2- and 4-door forms, and its looks underlined the abilities of the Turin based Pininfarina styling house. The clever packaging and pleasing lines were achieved at the expense of the Morris having a small, shallow boot which compared adversely with the Cortina's roomy 21 cu. ft carrying capacity. This shortcoming did not prevent the 1100 becoming Britain's number one family saloon as from 1963, and until 1972 it was the country's best selling car—though the Cortina Mark II edged ahead briefly in 1967. By 1973 Ford had introduced two new Cortina models and only then finally outsold the dated 11 year old front-wheel-drive car. Ford had got in front and stayed there.

In 1965 the 1100's sales contributed to BMC achieving a 34 per cent market penetration, the highest output

The Cortina goes into production at Ford's Dagenham body plant (above and left) *with 2- and 4-door versions being produced concurrently*

Above *The finished product, 'the medium sized car at a small car price', all set to take the British market by storm. This 1962 De Luxe 2-door Cortina cost £666 which was £27 more than the Standard version. Extras on this example are whitewall tyres (£6 14s 9d) and two tone paintwork (£6 17s 6d)*

Far right *Left-hand drive Cortinas all set for export pictured on Ford's Dagenham jetty in 1963. Italy was Ford's largest European export market at the time*

then achieved by the combine. BMC then merged with Jaguar in 1966 to become British Motor Holdings. However, following two years of falling profits, BMC was effectively taken over by the expansive Leyland Motors to form the British Leyland Motor Corporation in 1968.

Sadly it was the Mini story all over again: the technically advanced front-wheel-drive cars had been drastically underpriced suggesting that the Corporation's management did not have a strong enough grasp or perception of its own manufacturing

A line up of Standard Cortinas, identifiable by their painted radiator grilles and lack of brightwork around the headlights. These were just some of the 300 purchased by the London based Welbeck Motors, a mini-cab operator. The date is March 1963

102 FLU
101 FLU

Right *The Super's instrument panel was basically similar to the 1200 model though refinements included a carpeted floor and fresh air heater, the latter being an optional extra on the 1200 and costing £17 3s 9d*

Left *Interior of the 2-door Super with carpeted floor and different patterned seats and side trims*

Below left *The 1500 cc Cortina Super announced in January 1963, identifiable by its highlighted side stripe and aluminium wheel trims. Price was £688, £22 more than the 1200 De Luxe*

Far right *Great rival to the Cortina was BMC's Morris 1100, announced in August 1962, a month before the Ford. Its transverse engine and front-wheel-drive perpetuated the Mini theme though the interconnected independent Hydrolastic suspension and front disc brakes were new. It remained Britain's best selling car for close on a decade but failed to generate sufficient profits for its manufacturer*

costs. The ageing plant on which some of the cars were built contributed to an equation which spelt out financial disaster for BMC. Yet paradoxically Ford probably had a greater awareness of its rival's manufacturing costs than the Corporation itself! Frank Harris, who took over from Terry Beckett as Ford's Product Planning chief in 1963, initiated a searching and comprehensive study of BMC's engineering and financial strategy. Harris came to conclusion that the combine was heading for bankruptcy. On the report's completion Ford's Sir Patrick Hennessy telephoned George Harriman, his BMC opposite number who had taken over the chairmanship in 1961. Hennessy informed him of the inescapable conclusion that the Corporation was underpricing its products which could only result in a continual erosion of profits. There is no evidence to suggest that Harriman heeded these warnings.

Possibly this reflected BMC's attitude to Dagenham at the time. This was well illustrated when a member of Ford's Product Planning staff rang the chairman's office to offer a Cortina for appraisal. (This reciprocal arrangement was operated then, as now, by manufacturers to evaluate the other's products). However on this occasion the offer was brushed aside with the reply that the BMC chairman was not interested in what Ford was doing. . . . Inevitably, history has had the last word.

For even though the 1100 was successful on the home market the front-wheel-drive car from Cowley was strongly challenged by the Cortina. Due to Ford's successful export penetration, around 300,000 were manufactured within the first year of production which constituted a record British output. The millionth Cortina, built in September 1966, just four years after its announcement was another first, and by the time production ceased later in the year a total of 1,010,090 of these cars had been made. It was then replaced by the Mark II model. The Ford Cortina was well established and destined to become an integral part of the British

562 MWL

First facelift; the Cortina's original but very cheap ribbon speedometer was replaced by twin dials for the 1964 model year. However, the two spoke steering wheel was perpetuated. Automatic came too

motoring scene until it finally ceased production in Mark IV form in July 1982. The end was just over 20 years after the first models left the Dagenham production line. The meticulous costing approach that had characterized the Archbishop project was continued in the Mark I's successors. In recent years this has been a major contribution in ensuring that Ford has remained Britain's only major profitable car manufacturer.

The respective Cortina models have benefited from continual refinement and improvement, perpetuating a

policy initiated on the Mark I cars. The 2-door Standard and De Luxe forms were announced in September 1962, and a 4-door version quickly followed. This was unveiled at that year's Motor Show when Ford took its largest ever stand and displayed no less than five Cortinas. In January 1963 the 1489 cc Super was announced, instantly identifiable by its side flash accentuated by chrome trim, bright metal window frames and aluminium wheel trims. The model's top speed was consequently upped from the mid-70s to 80 mph plus.

MWC 879C

Left *The Super, refined for 1965. The* Cortina *legend replaces the* Consul *one on the bonnet and there is a new, two tier radiator grille that incorporates the side lights. The Aeroflow outlet grille can be seen behind the rear side window and the triangular badge on the rear wing indicates Super status*

Above *The new, greatly improved dashboard layout for 1965 with recessed dials, Aeroflow 'fisheye' outlets and a more substantial three spoke steering wheel. This is the column gear change version*

Overleaf *Rear view of a 1965 1200, a sight familiar to many motorists in the early sixties*

GB
Cortina
FOO 836B

Right *The Cortina's 21 cu. ft boot which was considerably larger than its Morris 1100 rival. The outline of the top of the petrol tank which did double duty as the floor, can be clearly seen*

Below *The 1965 Estate version having received its new radiator grille and Aeroflow outlets. This is the De Luxe version*

Weight increased from 15.5 cwt to 16.4 cwt, so there were 9 in. rather than 8 in. drum brakes, and 5.60 × 13 tyres in place of the 1200's 5.20 covers. Inside the car there were two colour seats and a standardized heater, the latter being an optional extra on the 1200 version. The week after the Super's announcement came news of the Lotus Elan engined Lotus Cortina. This and the GT version are considered in the next chapter.

Estate versions were revealed in March at the Geneva Motor Show. The Super sported simulated wood *Di-Noc* panels along its sides and rear which some regarded as tasteless adornment. However the stylist, Roy Brown, was enthusiastic about this American established process. It was 1500 powered, though 1200 and 1500 options were available on the cheaper and chromed trimmed De Luxe. As if to echo Lotus Cortina practice, a

Nice lines; the 2-door Super of 1965. The simple yet functional styling has well stood the test of time

The Rt Hon. Roy Jenkins (centre) as Minister of Aviation in Harold Wilson's 1964 Labour government makes the acquaintance of the Cortina Super in America. To his left is Ron Platt, responsible for British Ford sales in the United States

contrasting side stripe was available at extra cost.

The 1963 Motor Show at last gave Ford the opportunity to give the Cortina a much improved instrument panel. The simple, low cost strip speedometer was replaced by the more substantial dial type, in the manner of the Lotus Cortina. Other changes for the 1964 model year included the elimination of grease nipples from July 1963 and 'silent shut' child proof locks from September.

In January 1964 came a Borg-Warner Type 35 automatic gearbox option for £82 extra. However the really big improvements came with the 1965 Cortina, revealed at the 1964 Motor Show. Superficially the cars looked similar to their predecessors but there was a new, redesigned radiator grille that incorporated the combined side lights and flasher units. Also the *Consul* legend on the dummy bonnet air intake was replaced by one simply reading *Cortina* as the model was henceforth to be known. The only other major external differences were grilles which were introduced behind the side windows, revealing the presence of a new heating and ventilation system. The installation of this Aeroflow facility required the introduction of a greatly improved

It didn't look like this when it left Dagenham! Based on the Cortina GT and finished in Borneo Green this car appeared on Harold Radford's stand at the 1964 London Motor Show built to Stirling Moss's specifications as denoted by the registration number. Extras included Motorola radio, tape recorder and electrically operated side windows. It was priced at £1390 which was £160 more than the GT

fascia with new, so called fish eye adjustable outlets at its extremities. Thus ventilation could be achieved without the driver opening the car's windows though the by now superfluous front quarter lights were retained. In addition, improved seats were introduced. The original two spoked steering wheel was replaced by a more substantial three spoked one. On the mechanical front, 9½ in. disc brakes were introduced on the front wheels of the 1200 model. The aforementioned improvements resulted in across the board price increases, for instance the 4-door Super version cost £702 instead of £689.

Changes for 1966, the last year of the Mark I Cortina's production, included the discontinuation of the steering column gearchange option also with the no frills Standard model, though a version was continued in CKD form for the Dutch market. The Mark I Cortina

Above *Cortina derivative; also styled by Roy Brown with more than a reminder of the Taunus 17M, the Consul Corsair was introduced for 1964 to replace the Classic. Based on a lengthened Cortina floor pan and thus sharing its track, the Corsair was 1500 powered though was V4 engined from 1965. Output continued until 1970*

Far right *A Cortina on home ground. The Ford rally team visited the Italian Alpine resort of Cortina d'Ampezzo in December 1964, Terence Beckett having named the car after the venue of the 1960 Winter Olympics*

was dropped in the autumn of 1966 to make way for the Mark II version. The latter shared its predecessor's floor pan and was to prove just as successful.

It was just this success that paradoxically resulted in Ford scoring an 'own goal' as far as the Classic was concerned. Introduced in 1961 it was replaced by the Cortina derived Corsair for 1964 after 128,206 models of it and its sporty Capri variant had been produced. Terence Beckett is characteristically candid about the Classic's demise: 'Where we slipped up in a way was delaying its introduction by 12 months, but it was a carry over from a previous strategy: over designed and overweight. One of the arts of management is that if you recognize you've done something like this is not to endlessly cry about and try and keep it going but drop it and move on'.

CORTINA
SALUTE
TO CORTINA
CHAMPIONS
DECEMBER
2nd.–3rd.
1964

CHAPTER FOUR

Faster: Lotus and GT

Ford of Britain's involvement with motor sport in the fifties was essentially a low key affair with the competitions department, such as it was, based at Lincoln Cars premises on the Great West Road. It was true there had been a win for Maurice Gatsonides at the wheel of an aluminium bodied Zephyr in the 1953 Monte Carlo Rally, and there had been other sporadic and sometimes surprising successes. However, company participation increased with the arrival of the 105E Anglia with its light, over square engine.

As chronicled earlier, the development of the Archbishop project had resulted in a strong and above all light car, with the impetus on keeping the car's selling price as low as possible. Sir Patrick Hennessy and Terry Beckett had shown the car to Colin Chapman in the spring of 1962 and suggested that his firm, Lotus, might like to produce a sports racing version of the Cortina suitable for Group Two homologation.

Chapman had been building sports car since 1953 from cramped premises in Tottenham Lane, north London, offering them mostly in kit form. Although a competition programme was initiated, Chapman's first significant road-going car was the Lotus Elite coupé. In 1957 this was the world's first fibreglass monocoque, superbly styled by his accountant friend Peter Kirwan-Taylor. Although the Coventry Climax engined Elite was a sensational good looker, it was noisy, lacked refinement and more seriously, Lotus was losing around £100 on every one it built. A new sports car was therefore a

The Lotus Cortina, announced in January 1962 and pictured, appropriately, at Brands Hatch in 1963. The quarter bumpers, distinctive white finish with green side stripe and lowered suspension are all too apparent. The Lotus badge on the radiator and rear wing also identify

Right *What the other drivers usually saw, the rear view of a Lotus Cortina. The boot lid is aluminium and bereft of the usual* Cortina *legend. The model was only produced in 2-door form*

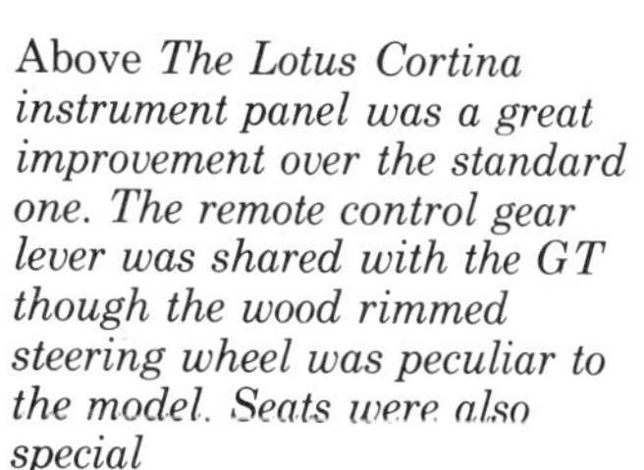

Above *The Lotus Cortina instrument panel was a great improvement over the standard one. The remote control gear lever was shared with the GT though the wood rimmed steering wheel was peculiar to the model. Seats were also special*

Far right *The twin overhead camshaft Elan engine which powered the Lotus Cortina undergoing adjustment at the Lotus factory at Delamare Road, Cheshunt, Herts in 1963. Note the vacuum brake servo, a standard fitment on the model*

priority and Chapman embarked on a design that would permit open bodywork, an impossible option on the fixed head Elite. So the new car was built up around a backbone chassis with all-independent suspension; it was then a matter of deciding on a suitable power unit. Colin Chapman was against the costly Coventry Climax engine, for by 1958 he had been thinking about developing his own unit. Initially the 4-cylinder Ford Consul engine, complete with Raymond Mays Conversion was considered. However 1959 saw the arrival of the 105E and Chapman recognized that its engine would form an excellent basis for a unit to power the new car. Harry Mundy, who had worked for BRM and Coventry Climax prior to becoming *The Autocar*'s technical editor in 1955, was commissioned to design an aluminium twin overhead camshaft cylinder head that would be mated to the 105E's cylinder block. Fortunately for Chapman, Mundy had already completed the preliminaries for a twin cam engine for use in the Facel Vega Facellia sports car, although the project had been aborted. Using this as a starting point, Mundy produced the design though it was not used with the 105E block but the 1340 cc Classic engine that had appeared in May 1961. This first experimental engine ran on October 16th 1961. However even before this version could properly evolve Ford brought out yet a further 1498 cc derivative which powered the Classic from 1962.

By 1960 Ford in Britain was considering an entry into the cheap sports car market and planning to manufacture the new Lotus sports car. The intention was to market it under the Lotus Ford name and for it to be steel rather than fibreglass bodied produced on cheap Kirksite tooling, limiting output to around 100 cars a day. However, the plan was scrapped following a temporary collapse on the American sports car market in 1959. So Chapman decided to press ahead and produce the car from the new factory in Cheshunt, Herts where Lotus had moved to in June 1959. Yet a further casualty was a Ford-designed vehicle—Philip Ives, who was

Above *The greatly improved Lotus Cortina dashboard with Aeroflow heating and ventilation. Note the Lotus monogram on the centre of the steering wheel*

Right *The Lotus Cortina received the updates of the Dagenham cars for 1965 with new radiator grille,* Cortina *bonnet badge and Aeroflow outlet identifying*

Overleaf *A police force takes delivery of five Lotus Cortinas in 1965. These cars purchased by the West Sussex Constabulary were standard models though with alternators in place of the original dynamos. They are also lacking their green side stripes, so that offending motorists might think they were standard cars!*

CORTINA
GWC 252B

POLICE
POLICE

POLICE
POLICE
GBP 3C
GBP 2C

LOTUS
LOTUS

given the task of undertaking the feasibility study in 1960, recalls that an MGA, MG Midget, Sunbeam Alpine and Austin Healey 3000 were purchased for evaluation. A design was conceived, but the Ford Product Planners were amazed to find that their dimensions almost exactly coincided with those of the 105E powered Berkeley Bandit, announced in 1960. This model is greatly to the credit of the small caravan manufacturer from Biggleswade, Bedfordshire.

Meanwhile work was progressing at Lotus on the development of the 1498 cc Classic based twin cam engine for the new sports car, though some modifications were necessary to the camshaft drive in view of the block's slightly greater height. The new engine was then fitted in a Lotus 23 and entered in the Nurburgring 1000 km race in May 1962 driven by Jimmy Clark. Although the car only lasted five laps before its exhaust flanges broke off and the resulting fumes forced Clark to retire, he put up the fastest lap in the two litre class which was a great encouragement for Chapman to press ahead with the engine's production. However following Ford's approach it was found that the twin cam could be used in two cars rather than one, since in addition to its employment in the Elan it could also power a sports racing version of the Cortina.

By the early sixties Ford in Britain, like its American parent, was becoming increasingly aware of the importance of the performance image to woo the younger car buyer. An added stimulus came from BMC who had gained valuable publicity by the competitive successes of its Mini-Cooper and the use of a car that outwardly, at least, resembled the production version made sound marketing sense. Ford were also acutely aware that while the Cortina met all the requirements of a family saloon it was a rather ordinary car which lacked sporting appeal.

So in June 1962 Ford's Product Planning Committee requested that Philip Ives investigate the possibility of improving the performance of the 1000 cc Anglias that at

Far left *For 1966, the final year of the Lotus Cortina's racing career, the cars ran to Group Five specifications with fuel injected, dry sump engines. As can be seen by the plate on the timing chain cover, BRM was responsible for preparing these units which were capable of up to 180 bhp*

Henry Ford at the wheel of the prototype Cortina GT at Montlhéry in June 1962. Fred Hart is occupying the passenger's seat

the time were being outpaced by the Mini-Coopers. He was also to carry out an investigation on the Cortina to establish the best means of enhancing Ford's performance image.

The following month the committee agreed that the 1500 Cortina, due to be announced the following year, would be produced in three stages of tune:-

Tune 1 A 79 bhp Dagenham designed, developed and manufactured version for sale as an improved touring car analagous to the Vauxhall VX 4/90
Tune 3 A 125 bhp version designed, developed and produced by Lotus for entry by Ford in Production Car races under Lotus's management.
Tune 2 A 100 bhp detuned derivative of Tune 3 to be produced by Lotus for sale through Ford dealers. A minimum of 1000 would have to be produced within 12 months to qualify for RAC homologation.

After driving the car Henry Ford sanctioned its manufacture and even ordered an example for his own use, much to Fred Hart's delight. The 1500 GT badge was only fitted to this prototype

The Tune 1 version of the tripartite package was destined to emerge as the Cortina GT, while the other two were track and road versions respectively. The latter are what we now know as the Lotus Cortina, though the less popular but correct rendering is Cortina Lotus. Designated the Lotus Type 28, it was revealed to the public in January 1963. Externally the car was unmistakably a 2-door Cortina, white finished, with the all-important side stripe picked out in contrasting green. The bodyshell was considerably lightened by the fitment of aluminium doors, bonnet and boot lid, and by 6.00 × 13 rather than 5.20 tyres. Under the bonnet was the 1588 cc Elan engine developing 105 bhp in standard form, only about 50 of the sports cars having been built in 1498 cc form. Lotus had abandoned Ford's sacred 80.96 mm bore and increased it to 82.6 mm, though the stroke remained at 72.75 mm. In the interests of weight

Visually identical to the Cortina De Luxe, the GT was only identifiable by the discreet badges on its rear wings. This 2-door version cost £748 on its 1963 announcement, £79 more than the Super. The 4-door GT cost £766

GT

The 1500 GT engine with twin choke Weber carburettor, water cooled inlet manifold and full flow exhaust system

saving, the clutch housing, gearbox tailshaft and differential housing were of aluminium. There was a radical departure from the original leaf spring layout at the rear of the car, for the axle was located by an A bracket mounted directly beneath the differential housing, the arms of which spread eagled almost the entire width of the car. Coils replaced the leafs as the suspension medium. In addition there were twin trailing arms attached at one end to the shock absorber mountings on the axle and to the forward spring hanger on the other. There was additional structural reinforcement inside the car's boot.

At the front of the Lotus Cortina, the MacPherson Struts were lowered and stiffened. A forged track control arm replaced the pressing on the production Cortina and was also slightly longer to eliminate wheel camber. In addition the anti roll bar was stiffened up and extended by ½ in. each side to reduce castor action. Shorter

Sectioned for exhibition purposes, the GT's engine was the 1500 unit with five bearing crankshaft, the latter feature being considered essential for this larger capacity

steering arms were introduced to gear up the radius. The servo assisted brakes were to be shared with the as yet unannounced GT, these were $9\frac{1}{2}$ in. discs at the front and 9 in. rear drums. The car's black finished interior was a great improvement over the cost conscious 1200 Cortina; with improved seating, remote control gear change, wood rimmed steering wheel and a new instrument layout dominated by a large circular rev counter and speedometer. Turning the scales at 16.5 cwt, the car weighed only slightly more than the Cortina Super. The judicious weight savings had to some extent been offset by the larger cylinder head, twin DCOE Weber carburettors and bigger wheels. Top speed was 103 mph and the car sold for £1100.

Interior of the GT with steering column mounted rev counter, looking very much an afterthought as indeed it was. The remote control gear lever was shared with the Lotus Cortina though the console mounted oil pressure gauge and ammeter were only fitted to the GT

Lotus produced the car from 1963 until 1966, during which period a number of modifications were made to the original design. The first major change came in July 1964 when the aluminium body panels were replaced by steel, though there was a mixture of the two during the change over period. Similarly the other aluminium castings were replaced by steel production parts. There had also been complaints by drivers of torsional vibrations between 55 and 70 mph so the original propellor shaft was replaced by a split unit. At the same time the gearbox, which was an Elan close ratio one, was replaced by a Cortina unit with an uprated second gear.

Above *Rear of the 1963 GT. Seats were in single colour PVC and the cars carpeted throughout*

Overleaf *The Cortina was soon earning its spurs; Henry Taylor and Brian Melia in a 1200, running to Group Two specifications, in the 1963 Monte Carlo Rally where they finished second in class.* Inset, *the same car in a less demanding competitive event*

The 1965 model year saw the Lotus Cortina benefit from the standard car's updates, namely an Aeroflow heating/ventilation system and revamped instrument panel. The latter of course, differed from the basic Cortina and GT.

Then, in July 1965, the Chapman modified rear suspension was discarded as it had suffered occasional collapse through the differential leaking oil and the consequent deterioration of the A bracket's bushes. It was also more vulnerable to damage on boulder strewn rally courses. In its place came the simpler half elliptic sprung rear axle with twin radius arms as on the GT.

193
TOO 52

41
TOO 528
193

Above *An early Cortina all set for the competitive fray, well equipped for a night time event and tricky terrain!*

The last major alteration to the design came in October 1965 when the gear ratios were again changed and replaced by those used in the Corsair 2000E. This had a much higher first gear and slightly higher second and third cogs. At the same time the brakes were modified with the front caliper moved from the rear to the front of the disc, and self-adjusting back brakes were introduced. Also GT seats replaced the specially commissioned Lotus ones. In short the car was becoming less of a special and more of a standard product. However these changes went in tandem with greater reliability and by the time that the model was discontinued in the autumn of 1966 the Lotus Cortina was a proven, potent and, above all, consistent performer. It should not be forgotten though that the whole reason for productionizing the car was so it could qualify for Group Two homologation. But before chronicling the model's outstanding competition record it is important to consider the conception and evolution of the Tune 1 Cortina option approved in June 1962; the sporting though less potent GT model.

As we've seen, the need for a sporting image had already been exercising corporate thought since around 1960 which made a sporting version of the Cortina desirable. However the GT began unofficially, a most unusual state of affairs at Dagenham, in response to news of the Lotus Cortina, Fred Hart remembers. Its starting point was a robust version of the Cortina which was being developed for Ford Australia. Hamish Orr-Ewing recalls that it had to be capable of being driven 'flat out over bad surfaces', in other words Australian driving style. This demanded a more robust sheet metal structure and suspension. Therefore rather than tamper with the existing weight conscious bodyshell, a heavier version was designed in parallel with it. So it was this derivative that formed the basis of the GT, and a prototype was created using this structure. 'All this was most irregular since such "have a go" ventures were very much frowned on in Ford engineering circles where nothing was made in metal until released from the drawing office' remembers Orr-Ewing.

He was a keen motor sport enthusiast and had approached Ford's engineering department with a view to their 'tweaking' the 1498 cc version of the versatile four, but such tampering with an existing specification flew in the face of their sensibilities. So he talked with Philip Ives who shared his enthusiasm for competitive motoring, and Ives went off to see Keith Duckworth of Cosworth Engineering. The North London based Cosworth company had been established in 1958 by Mike Costin, who was technical director of Lotus at the time and Duckworth who had just left the company. From 1960 onwards the firm had rapidly built up a

Overleaf *Pat Moss and Ann Riley in their Cortina 1500 during the 1963 East African Safari event. Unfortunately they retired after over turning on the Mubulu Plateau*

KHL 487

reputation for modifying the 105E Anglia engine for Formula Junior racing, and were also responsible for developing high performance versions of the Elan twin cam.

Cosworth came up with an improved 1498 cc unit which pushed up the engine bhp figure from 59 to 78 by changes to its specification. These were to result in the minimum of interference to the Dagenham production line. Larger inlet valves were fitted and a Cosworth developed high lift camshaft introduced. Another minor head modification was a relieved edge on its inlet side to improve gas flow. It featured a twin choke Weber carburettor, making the Cortina GT, together with the Capri equivalent which preceeded it, the first British production cars to be fitted with the unit. The compression ratio was 9.1 compared with 8.3 for the standard 1500, therefore stronger pistons and copper lead shells were used; the latter in place of white metal ones. Stronger clutch springs were fitted and the propellor shaft diameter increased.

With an engine in their possession, the Product Planners then handed it over to engineering 'who somewhat reluctantly agreed to test it and they discovered it hadn't got resonance or undesirable boom or valve bounce' remembers Ives. It was then a matter of going to the manufacturing department for a special exhaust manifold and the like. Other modifications to the car proper included a stiffening up of the suspension, and the introduction of 9½ in. disc brakes on the front wheels and 9 in. drums on the rear; by contrast the 1500 Super had drums all round.

If the method of the GT's conception was unusual then the manner in which it entered production was equally unconventional. Somewhat unofficially, an example was taken to the Cortina/Taunus 12M launch at Montlhéry, and it was there that Henry Ford had an opportunity of driving the car where it won his immediate approval. However the use of the GT name caused some heart searching. 'A few purists amongst Ford's management

The GT's instrument panel as restyled for 1965 with auxiliary instruments and Aeroflow heating and ventilation

considered it pretentious' says Orr-Ewing, but with the approval of the specification work went ahead in preparing the car for production.

In addition the engine, suspension and brake modifications were used on the Classic based Capri GT model which was announced in March 1962, a month prior to the Cortina version. Its interior changes included the introduction of a remote control gear lever, as in the Lotus Cortina, though less satisfactory, attached to the steering column was a hooded rev counter looking very much an afterthought. An ammeter and oil pressure gauge were added to the central console ahead of the gear lever. Interior trim was based on the Cortina De Luxe. The only special external indication about the 90 mph model was the presence of

GT badges mounted on each rear wing. In view of the GT's antipodean specifications the car was 1 cwt heavier than the Super, also 1500 cc engined, at 17.5 cwt.

The GT received a new instrument panel, along with the other Cortina models for the 1964 model year, though the rev counter was peculiar to the variant. There were further changes at the 1965 Motor Show when it benefitted from yet another dashboard revision demanded by the fitment of the Aeroflow heating and ventilation system. This included the introduction of four centrally mounted auxiliary instruments: a great improvement. Less obvious were the competition proven twin radius arms added to the rear axle. There were no further major modifications made to the model before

Roger Clark at the wheel of his red GT Cortina stopping at a check point on the 1965 Scottish Rally he later went on to win

the Mark I Cortina range was phased out in the autumn of 1966 to make way for the Mark II model which perpetuated the GT theme.

But what of Ford's all-important competitive ambitions? 1963 represented a turning point in the history of the company's competitions department, which up until then had operated on a consistent but rather low key basis. The Lotus and GT Cortina were announced that year, and with a move to Boreham Airfield in Essex, Ford's commitment to motor sport took a significant and positive leap forward. Boreham's sphere of activities would be confined to preparing GTs for rallying while the racing programme, spearheaded by the Lotus Cortina, would operate under the auspices of Team

The late Jim Clark (left) sharing the bonnet of his Lotus Cortina with Roger Clark during a rally practice session in Kent during 1966. The car was subsequently written off by Roger

Far right *As the Lotus Cortina is often remembered; Bengt Södeström taking to the air at the wheel of his left-hand drive car with Gunnar Palm in the passenger's seat during the 1966 RAC Rally that he later won*

Lotus and Alan Mann Racing. Initially Syd Henson was appointed competitions manager and was succeeded by Alan Platt and Henry Taylor respectively, during the competitive life of the Mark I Cortina.

Not surprisingly the less sophisticated Cortina GT was first in the field, and Pat Moss got off to a fine start at the Tulip Rally in May. This followed the model's announcement in April, when they were awarded the Coupes des Dames prize with Jennifer Nadin. In the same month Henry Taylor achieved a fourth overall placing in the Acropolis Rally when GTs also chalked up first, second, third and fourth class placings. Taylor followed this up with a class win in the famous Alpine Rally in June. The punishing Spa-Sofia-Liege event in August saw Taylor and Brian Melia attain fourth place in a car that was effectively a Lotus Cortina but with leaf spring rear suspension. The Taylor/Melia duo scored a sixth overall placing in the RAC Rally in November, while Pat Moss in her GT carried off the Ladies Prize and Ford received the manufacturers team award. It had been a good first year.

1964 saw the Lotus Cortina begin to show its mettle on the race track. However the first really impressive victory for the company came in a road event when Cortina GTs swept the board in the gruelling East African Rally held in March. The Peter Hughes/Bill Young car achieved an outright win with Armstrong and Bates in third place. It had only been Erik Carlsson and his Saab that had prevented a Ford hattrick, though they won the manufacturers' team prize and took the first four class placings. It is also worth mentioning that Ford achieved further successes in the same event in 1967, *after* the Mark I Cortina had ceased production. Lotus and GT triumphantly upheld Ford laurels with the cars taking second and third placings respectively along with the team prize.

The competitive spotlight now turns on the twin cam Lotus Cortina. There was a victory for two Alan Mann entered cars in the *Motor* Six Hour International

NVW239C

Touring Car Race at Brands Hatch in June. Here they received the chequered flag ahead of two works entered Mercedes. Following this, at the Oulton Park Gold Cup meeting in September, Jack Sears and Trevor Taylor managed third and fourth placings behind a pair of Ford Galaxies.

In 1965 the twin cam cars really got into their stride with Sir John Whitmore winning the European Touring Car Championship in a leaf sprung Lotus Cortina. Other rally successes included Jackie Ickx winning the Belgian Saloon Car Championship, and Bengt Söderström attaining Swedish Rally victory laurels. Privateer Roger Clark won his second successive Scottish Rally in a GT Cortina and in December, his first event as a member of the Ford works team when he chalked up a win in the Welsh Rally at the wheel of the Lotus Cortina. Not that competition successes were confined to the European theatre. A Lotus Cortina won the New Zealand Gold Star saloon car championship in 1965 while a GT was the Canadian Winter Rally winner; truly an international competition car.

There were differences then in the mechanical specifications of the Lotus Cortinas from those that ran in 1966. The cars competing in Group Five form had 180 bhp BRM tuned Lucas fuel injected dry sump engines and the MacPherson strut suspension, these were replaced by a coil spring and damper layout. That year's victories included the Acropolis and Geneva rallies, the Lotus Cortina won two coupés des alpes in that year's Alpine event. The model also won the 1966 RAC Rally when Bengt Soderstrom and Gunnar Palm took a left-hand drive car to victory. This followed major wins in the Shell 4000, Acropolis and Geneva events. The team also had its share of bad luck during the year and were robbed of victory in the Italian Rally of the Flowers on a technicality. However in 1966 the original Cortina made way for its Mark II successor and Ford's Boreham based competitions department continued to build on the success achieved in those three formative years.

Specifications

	Engine, transmission and final drive ratios
1200	80.96 × 58.17 mm 4-cylinder, 1198 cc developing 48.5 bhp (net) at 4800 rpm. Maximum torque 63.5 lb ft (net) at 2700 rpm. Compression ratio 8.7:1, optional 7.3:1. 1965 onwards 9:1, Solex B30 carburettor. Cast iron block/crankcase, cast iron cylinder head. In line vertical overhead valves operated by tappets, pushrods and rockers from chain driven camshaft. Skew drive from camshaft to distributor and external rotor type oil pump and AC filtre. 7.25 in. single plate clutch, 4-speed all synchromesh gearbox. Overall ratios, top 4.13, third 5.83, second 9.88, first 14.62, reverse 16.35. Final drive ratio 4.125:1
1500	As above but 80.96 × 72.75 mm, 1498 cc developing 59.5 bhp (net) at 4600 rpm. Maximum torque 97 lb ft (net) at 3600 rpm. Zenith 33VN carburettor. 8.3:1 compression ratio, 7:1 optional. 1965 onwards 9:1, Overall gear ratios, top 3.9, third 5.5, second 9.34, first 13.8, reverse 15.5. Final drive ratio, 3.9:1, Estate 4.44:1 (Super and 1500 De Luxe 3.9 or 4.1:1)
GT	80.96 × 72.75 mm, 1498 cc developing 78 bhp (net) at 5200 rpm. 28/36 DCD 16/18 twin choke Weber carburettor. 1965 onwards 28/36 DCD 22, 9:1 compression ratio
Lotus Cortina	82.6 × 72.75 mm, 1558 cc developing 105 bhp (net) at 5500 rpm. Compression ratio 9.5:1. Twin Type 40 DCOE/18 carburettors. Chain driven twin overhead camshafts in cast aluminium cylinder head. 8 in. diaphragm clutch. Overall gearbox ratios, top 3.9:1, third 4.8, second 6.4, first 9.76, reverse 10.959:1. Final drive ratios 3.9:1 or optional 3.77, 4.1, 4.43. From July 1964 3.9, 5.51, 7.96 and 13.82, reverse 15.46. From October 1965, 3.9, 5.45, 7.84, 11.59, reverse 12.95:1.
Steering	Burman recirculating ball
Suspension	Front, MacPherson struts and Armstrong telescopic dampers, and anti roll bar. Rear, live axle with leaf springs and Armstrong telescopic shock absorbers

GT	From October 1964, radius arms introduced on rear axle.
Lotus Cortina	Rear, live axle, located by radius arms and A frame. Coil springs mounted coaxially with dampers. From June 1965 GT type with half elliptic springs and radius arms.
Brakes 1200	Girling hydraulic. Front 8 in. drums, 1.75 in. shoes. Rear 8 in. drums, 1.50 in. shoes.
1500	Front, 9 in. drums, 1.75 in. shoes. Rear 8 in. drums, 1.5 in. shoes.
GT	Front, 9.5 in. discs. Rear 9 in. drums, 1.75 in. shoes
Lotus Cortina	As GT but with vacuum servo assistance.
Wheels and tyres	Pressed steel four stud wheels, 5.20–13 tyres
Super	5.60–13
Estate	6.00–13
Lotus Cortina	6.00–13
Dimensions	Wheelbase 98.25 in. Front track 49.5 in. Rear track 49.5 in. Overall length 170.5 in. Overall width 63 in. Overall height 57 in.
Weights	Kerb weight (lb) 1200 2-door De Luxe 1775 1200 4-door De Luxe 1803 1500 2-door Super 1863 1500 4-door Super 1899

Cortina type numbers and introduction dates

Type No. rhd/lhd	*Description*	*Engine*	*Introduction*
113E/114E	2-door Standard	1200	September 1962
113E/114E	4-door Standard	1200	October 1962
118E/119E	2-door Super	1500	January 1963
118E/119E	4-door Super	1500	January 1963
118E/119E	2-door GT	1500	April 1963
113E/114E	Estate De Luxe	1200	March 1963
118E/119E	Estate De Luxe	1500	March 1963
118E/119E	Estate Super	1500	March 1963
125E/	Lotus Cortina	1558	April 1963

Cortina production (1962–1966) with comparative Anglia (105E, 123E) and Classic/Capri (109E, 116E) output

	Anglia	*Classic/Capri*	*Cortina*
1959	48,139	-	-
1960	191,752	-	-
1961	146,028	51,416	-
1962	143,303	58,622	67,050
1963	110,184	17,161	264,332
1964	138,168	1007	221,678
1965	123,943	-	263,353
1966	111,463	-	193,677
TOTAL	1,083,955 *	128,206	1,010,090

* 70,975 Anglias built in 1967

Cortina Mark I and Taunus 12M: Comparative dimensions

	Cortina (Archbishop)	*Taunus (Cardinal)*
Wheelbase	98.25 in.	99.5 in.
Track	49.5 in.	49 in.
Overall length	171 in.	170.5 in.
Overall width	62.5 in.	63 in.
Overall height	56.25 in.	56.5 in.
Kerb weight (1200)	1792 lb	1863 lb

Mk1 Cortina Owner's Club

The Club was started in 1982 by two Mk1 Cortina enthusiasts from opposite ends of the U.K. From a nucleus of 25 members in May 1982, the club has grown to its 1994 size of some 1,200 members. The club now has members throughout the world, including Japan, USA, Australia, Sweden and Spain. The club is recognised by the Ford Motor Company, and caters for all Mk1 Cortina variants, whether standard or customised.

A friendly service is offered to owners and enthusiasts. The main aim is to locate spares and pass them on to clubmembers at very attractive rates. A bimonthly newsletter lists the latest spares available and is full of members' letters, stories and cars/parts for sale and wanted! The club re-manufactures parts no longer available and ensures that M.O.T. related items are always in stock.

The Mk1 is basically a strong and reliable car with any weaknesses known, together with the cure! Most technical advice and modifications are readily available from club sources.

The club has a stand at many of the classic car shows up and down the country during the season and encourages members to attend as many local shows as they can. There are also a number of informal "Pub Meets" held around the country which are run and attended by members on a monthly basic. Details of these meetings appear in the magazine.

Once a year the club holds a National Mk1 Cortina Day. The site has full camping/caravanning facilities so is ideal for those travelling long distances. At this gathering, which normally attracts some 150 to 200 Mk1 Cortinas, are other related Ford Clubs. The club try to have a celebrity associated with the Mk1 to present the prizes in the separate concours section, but your car need not be in "show" condition. All Mk1s are welcome, so please do make the effort as the amount of spares available will certainly make your journey worthwhile.

We hope you will decide to join the Club and thus help to perpetuate one of Ford's most successful motor cars. Contact the Membership Secretary, Roger Raisey at 51 Studley Rise, Trowbridge, Wiltshire, BA14 0PD, United Kingdom. The telephone number is 01225 763888.

Acknowledgements

I have had the utmost co-operation from many members of the team who created the Cortina but first and foremost my thanks must go to Sir Terence Beckett, Ford's Product Planning General Manager from 1955 to 1963. He not only provided a unique insight to the Archbishop project but also contributed the Foreword to this book. I have also received considerable help from Hamish Orr-Ewing, Light Car Product Planning Manager for the duration of the Cortina project while Fred Hart, formerly Executive Engineer Light Cars put me in the picture as far as the car's mechanical layout was concerned. Philip Ives provided valuable information relating to both the creation of the Cortina GT and to Ford's Product Planning department, as did Dennis Roberts in the vital body engineering field.
I am also greatly indebted to Ron Mellor, Vice President of Car Engineering for Ford of Europe and Ron Platt, Vice President of European Sales Operations for Ford of Europe; not only for their memories of the Cortina's creation but also for providing invaluable perspective material. Thanks are also due to Ford's archivist, David Burgess-Wise, who offered every assistance with the project.
Practically all the photographs are courtesy of the Ford Photographic Department, and my thanks to manager Steve Clark and to Sheila Knapman for providing prints with such speed and efficiency. Additional illustrations have been provided by *Autocar*, Fred Hart, Hamish Orr-Ewing and Ron Platt.

Index